Praise for

THE POTENTIALIST
THE PURSUIT OF WISDOM

"Wisdom is an invaluable, yet elusive, attribute in today's profoundly changing world, and its attainment is important for both individuals and society as a whole. Ben Lytle offers a guide for how each of us can become wiser, better people and citizens. It is indispensable reading for anyone who wants to fulfill their potential."

EVAN BAYH, attorney, advisor, board member, and former two-term governor and two-term US senator for Indiana

"I've known Ben Lytle for over twenty-five years as a CEO, serial entrepreneur, and creative thought leader. He has consistently pursued his potential, his family's potential, and the potential of the people and companies he led. He made the study of potential and wisdom his life's work. His insight into evolving social paradigms and how they affect us all is unique, yet simply expressed. His Potentialist series, and *The Potentialist: The Pursuit of Wisdom* especially, are gifts to all of us at the right time for the right reasons."

JEFF WEISS, founder of the CCI Executive Network, healthcare expert, and adjunct professor of medicine at the UCLA Department of Medicine

"Imagine possessing the knowledge, principles, and discernment of our wisest ancestors and the vision to flourish in the AI revolution with the highest ethical standards. This is an easy book to read and a hard one to put down; a wisdom handbook to guide you through the fourth industrial revolution that is now underway. I've known and deeply respected Ben Lytle for more than twenty years. He has had many successes, but in this opus are the principles and practices you need, and you will want to share them with the people you love."

ANNE RYDER, Emmy award-winning news anchor and documentarian, and senior journalism lecturer at Indiana University

"*The Potentialist: The Pursuit of Wisdom* is a must-read for anyone looking to increase their decision-making wisdom and unlock their full potential. In today's quickly changing world of technological advancements, Ben Lytle's groundbreaking work gives us the tools to adapt and thrive."

JOSH LINKNER, five-time tech entrepreneur, *New York Times* bestselling author, and venture capitalist

"In *The Potentialist: The Pursuit of Wisdom,* Ben Lytle issues a powerful call to unlock our inherent capacity for wisdom amid rapidly changing times. With clarity, he blends philosophy, psychology, and real-world experiences to reveal how we can live to our potential and find greater purpose. Drawing from his own journey of professional success, grace, and meaningful relationships, Ben shares invaluable insights garnered along the way. Get ready to feel inspired to nurture what lies within—it'll transform your life and positively impact those around you."

MITZI KROCKOVER, MD, internist, podcaster, investor, consultant on women's health, former founding medical director of the Iris Cantor-UCLA Women's Health Center, and VP of women's health at Humana, Inc.

"We are all constantly encouraged to 'live our best lives.' In *The Potentialist: The Pursuit of Wisdom*, Ben Lytle tells us exactly how to do it by discovering the wisdom, skills, and knowledge we already possess or can easily attain. Ben makes a compelling case for wisdom as the ideal adaptive response to a radically different twenty-first century, not for an elite few, but as a common ethos for eight billion people accelerating human progress."

WOODROW MYERS, MD, managing director of Myers Ventures, LLC; former medical director for Indiana, Ford Motor Company, and Wellpoint, Inc.; and Democratic Party nominee for governor of Indiana

"Not since J. Irwin Miller, thought leader and legendary CEO of Cummins Engine Corporation, has there been someone I've known with greater insight into wisdom than Ben Lytle. He explains how important it is, how difficult it is to define but easy to recognize, and how essential it is to our lives both now and in the future. *The Potentialist: The Pursuit of Wisdom* is a must-read for the coming decade."

JOHN MUTZ, business leader and public servant, lieutenant governor of Indiana, Republican candidate for governor, and president of Lilly Endowment

"For more than thirty years, I have observed the ultimate 'Potentialist'— Ben Lytle. This book embodies the nexus of intellectual capability, cognitive observation, and experiential optimization. Simply put, very few people can appreciate, integrate, and explain the layered complexity of the human personality, as demonstrated in this book. *The Potentialist: The Pursuit of Wisdom* clarifies the methods to achieve multiple successful life outcomes. A remarkable book, remarkable insights, and a tribute to a remarkable intellect."

JACQUE J. SOKOLOV, MD, investor in innovation, lifelong learner, and board chairman of multiple healthcare companies

"Ben Lytle challenges us to believe in our human potential and understand our internal and external worlds. He offers timeless questions and recommends steps to achieve lives dictated by our Souls and consciences and their worthy intentions. Ben shows the way to a life not dictated by ego, but by inner peace and reflection, to increase intimacy and wisdom."

JUDE MILLER BURKE, PhD, therapist, leadership coach, and author

"Much of what we hear and read today speaks of a frightening dystopian future. In *The Potentialist: The Pursuit of Wisdom*, Ben Lytle dispels this cognitive distortion, instead providing a vision of an opportunity-rich future replete with potential. Most importantly, Lytle guides readers, step-by-step, to realize their potential through cultivating wisdom."

SHERINE EMILY GABRIEL, MD, university professor and executive vice president for health at Arizona State University

"Having known and admired Ben Lytle for decades for his many professional achievements, I'm not surprised that he is tackling a very tough topic dating from antiquity—how to achieve wisdom. Ben offers an understandable theoretical framework, practical applications, and advice. His thorough and effective analysis of wisdom arrives at a time when it is more needed than ever!"

JEAN-PIERRE MILLON, healthcare expert, investor, board member, business advisor, and former Eli Lilley executive

"When I think of wise men and women, names like Socrates, Augustine, Frederick Douglass, Hypatia, and Mother Teresa jump to the forefront. None of these are known for commercial success. In volume two of his groundbreaking series, Ben Lytle guides us to developing the wisdom to navigate the shoals of life by reaching our greatest potential. It may not include wealth, but surely, it will be a life filled with meaning and impact. Read this book and be prepared to change your life and the world around you."

MICHAEL B. MURPHY, historian, author, communications executive, and political strategist

"In *The Potentialist: The Pursuit of Wisdom,* Ben Lytle has again crafted a practical approach for everyone to tap into their best self and the essence of living wisely and well. This insightful book is a toolbox of profound wisdom and practices for navigating life's challenges and seizing opportunities. Through thought-provoking discussions about the meaning of life, readers are empowered to unlock their potential and a purpose-driven existence. With a blend of philosophical reflections and practical advice, this book is a must-read for anyone seeking to cultivate a deeper understanding of themselves and the world around them."

CHRIS EPTING, *New York Times* bestselling memoirist and author of over forty-five books

www.amplifypublishinggroup.com

The Potentialist: The Pursuit of Wisdom

For more information, please contact:
Amplify Publishing, an imprint of Amplify Publishing Group
620 Herndon Parkway, Suite 220
Herndon, VA 20170
info@amplifypublishing.com

Library of Congress Control Number: 2024912393

CPSIA Code: PRFRE0824A

ISBN-13: 978-1-63755-719-8

Printed in Canada

This book is dedicated to you, the reader. You have selected it at a pivotal period in humanity's development. There will be more profound change in the next three decades than in the past several centuries. You are serious, or at least curious, about how to adapt, assist loved ones and others around you, and succeed during a time of radical change.

Most people think of preparing for the future as someone else's job, usually the public institutions of government, education, mass media, religion, and healthcare. You accept the responsibility that our collective destiny relies on each one of us.

Thank you for stepping up. The decades ahead can be the Age of Expanding Human Potential. You can lead yourself, your family, coworkers, and others in your orbit to their potential, wisdom, and success in the twenty-first century and be a beacon to others beyond. You're about to take the first step.

The Potentialist

The Pursuit of Wisdom

Ben Lytle

an imprint of Amplify Publishing Group

CONTENTS

FOREWORD

WHEN YOU THINK OF THE WORD *WISDOM*, what comes to mind? Please pause right here before reading on.

You are correct if you associate wisdom with words such as *thoughtful, considerate, authentic*, and *insightful*. If you connect it with advanced age, you may be surprised that wisdom is not limited to grandparents. Children as young as age four can possess nuggets of wisdom.

The fact that wisdom can be easily identified may be evidence that it is a trait and skill already residing within us. In this second book of The Potentialist series, Ben Lytle shows that it is possible to exercise the muscle of wisdom to create a better quality of life and relationships and help us reach our potential. You need not wait to be a grandparent to become wise!

As a licensed and board-certified clinical psychologist and a professor of psychiatry, I provided psychological consultation for Ben for nearly a decade as he developed concepts for this book. Originally intended only for his grandchildren, the book evolved to include Ben's half century of study of potential and wisdom, including insights, experiences, and findings from his voracious appetite for reading and learning and close associations with many wise people through conversations, interviews, and friendships. As a result, Ben has become an erudite lay scholar. Many concepts from psychology and philosophy tend to be full of dense and complicated jargon unfriendly to the general public. Ben translates the jargon into useful, meaningful, digestible concepts and practices. Applying Bloom's hierarchical taxonomy of learning often used by educators and clinicians, this book is at the apex of creativity, where Ben skillfully integrates complex thoughts of evaluation, analysis, application, and understanding. The book is a gem that predicts the near future and how we might adapt.

As a psychologist with thirty years of psychotherapy practice, I often use clinically appropriate books to supplement my work with clients and patients as part of their healing and growth treatment process. Before recommending them, I carefully consider factors such as the author's education and experience, required as part of providing ethical and competent care as a psychologist in accordance with the professional Code of Ethics. Although Ben does not hold an advanced degree in mental health, his reading and understanding are commensurate with those holding such advanced degrees. I am confident in vetting this book as a useful tool for therapists in their practices with patients seeking to deepen their insight and growth.

I truly believe wisdom is a gift from the Holy Spirit. It is our responsibility to develop that gift, which benefits not only ourselves but the whole of society.

—Lisa Colangelo Fischer, PhD, ABPP

PART I

Your Place in the World

CHAPTER 1

Potential, Wisdom, and the Art of Living Well

> "Life's tragedy is we get old too
> soon and wise too late."
> —BENJAMIN FRANKLIN

WHAT IF YOU COULD ACHIEVE YOUR POTENTIAL, become wise, make better decisions, have deeper relationships, and live well with only a small investment in time? Great news! You can, and this book will show you how.

The Potentialist: Your Future in the New Reality of the Next Thirty Years predicts accelerating change in turbulent, opportunity-rich times in the next thirty years as the last vestiges of Industrial Age structures and mindsets give way to the Age of Expanding Human Potential. The book proposes excellence in seven career and life skills that are the minimum wisdom you will need to adapt. *The Potentialist: The Pursuit of Wisdom* forecasts wisdom to become the twenty-first-century norm for success because the New Reality will demand it and make it possible for everyone willing to invest a little effort.

Your life can be richer and longer, and your contributions more consequential. As life expectancy expands, your peak period of creativity and wisdom can be thirty, forty, or fifty years. You will be part of a growing majority of older adults with historic financial and voting power and the wisdom to use them well. **The twenty-first century's greatest opportunity is mastering wisdom and the art of living well, which is demonstrated by wise decisions.**

This book is intended to aid the democratization of wisdom by simplifying the complexities of human potential development into accessible, contemporary, secular terms to make it understandable to more people. It also integrates futurist perspectives since adaptation to the New Reality may be the most significant human development challenge in history. This chapter summarizes what, why, and how to achieve your potential and wisdom. The following chapters provide the details and practices you will need.

WHAT ARE POTENTIAL AND WISDOM?

YOUR POTENTIAL IS THE BEST YOU

Inside everyone and every living thing is a natural drive to achieve their potential. That drive provides the forward momentum to maximize our individual life experiences while nurturing and contributing to the betterment of humanity, beginning with those closest to us.

Your potential is the realization of that drive; it is the best you, ready to be developed into your source of wisdom. Ben Franklin's truism on the previous page can be reversed. You and your loved ones can achieve potential and wisdom when you are younger and perform at your best over a longer life.

WISDOM IS THE ART OF LIVING WELL

It's such a simple but elegant truth. Wisdom has two components: the capacity for wisdom and the decisions that serve the decision-maker's interests and the interests of others. Capacity for wisdom means preparing ourselves to make wise decisions through knowledge and perspective of the world's reality and our place in it. Decisions are how we apply the capacity for wisdom to a specific factual situation. Your decisions, more than anything else, determine how well you live.

Wisdom is the pinnacle of human achievement and should be our highest priority. Yet, despite abundant information, role models, and experience, we are taught very little about wisdom or how to become wise. Throughout history, wise people's decisions, words, and deeds represent humanity at its best.

The wise are human, imperfect, and sometimes unwise. **A wise person makes better decisions more often, and their life reflects time-honored, admired qualities.** Wisdom requires preparation, including honoring timeless truths, knowledge of facts and circumstances, informed perspectives of life and the world, and freedom from Ego distortions, bias, and automatic responses. Wise people make mistakes just like everyone else, but substantially fewer of them. The only standard any of us can live up to is that we did our best with the capacity for wisdom we had at the time and independent judgment consistent with our times and cultures.

WHY PURSUE POTENTIAL AND WISDOM?

DEMOCRATIZATION OF POTENTIAL AND WISDOM WILL BE A NEW REALITY STEP CHANGE

Since the 1950s, success has been defined as having one or more college degrees, a professional career or business ownership, sophisticated knowledge of the world and current issues, and a worry-free retirement. Wisdom was considered exceptional and attributed to few people. This standard is expiring, and the bar for success is being raised. Achieving wisdom, exemplified by wise decision-making, will become the standard for twenty-first-century success. That may seem beyond the abilities of everyday people. Yet land ownership, literacy (owning and reading books), universal education, university degrees, and democratic government were once considered equally unrealistic. Expanding opportunities from a few to many and then to everyone are democratizing leaps forward, or step changes, in societal and individual development. Wisdom will be a democratization step change as the New Reality provides both the need for wisdom and the means to make it available to everyone.

CONVERGING CHANGE FORCES

The forces of change are fully explored in *The Potentialist: Your Future in the New Reality of the Next Thirty Years*. Following is a brief summary:

Demographic change

Birth rates have declined worldwide for over fifty years, while life expectancy has expanded. These trends will continue over the twenty-first century unless altered by cultural norms or technology that increases birth rates. Population growth will slow and decline starting with developed countries and then worldwide. More of the population will be older, wiser, and more prosperous. Each person's value will increase as there are fewer of us. Young people will continue to mature later as life expectancy expands and stages of life expand proportionately. Wisdom developed earlier and applied over a longer life will become increasingly valued.

Accelerating pace of change

Historically, life-altering change occurred more slowly, sometimes over centuries. In the last decade, the future arrived faster than anyone expected, even quicker than futurists and science fiction writers predicted. Our lives are changing faster than our physiology and psychology. Fear and anxiety accompany rapid change. Some people seek the fiction of certainty around extremes. That leads to polarization with a significant, silent middle ground trying to figure out how to respond. This is an archetypal pattern repeated many times in history. Challenging times will accelerate the demand for wisdom. Humans are adaptive creatures.

As in prior times of rapid change, a small percentage of the population will become New Reality adaptive beacons for others to follow. I hope you choose to become one, using this book as a guide.

Democratization

As technology-fueled democratization expands what has been rare to more and more people, individuals will adapt faster than societal institutions, cultures, and nations. Democratizing innovations have lifted individuals to greater abundance and equality over the last few centuries. In parallel, institutions have slowly declined in power and influence. As institutions provide less structure and fewer boundaries, individuals will assume greater accountability for their lives and success. Many complex decisions will be required. The best preparation for these decisions is for individuals to operate at their potential and maximum capacity for wisdom.

Deinstitutionalization

Democratization disrupts the established order. Corporations, industries, and public institutions will be disrupted, undergo radical change, or be deinstitutionalized and replaced. The New Reality will be volatile as insti-tutions become less unstable and unreliable before remerging in more adaptable democratized forms. We have witnessed the remaking of retail and entertainment and other industries and institutions. Education, representative democracy, mass communications, healthcare, commercial real estate, and banking are all likely to undergo significant change to adapt over the next one to two decades. Individuals will experiment with newfound freedom and power from depopulation and integration with technology.

Technology

Technologies will alter virtually everything we see today, especially medicine and the ubiquity of humanlike automation. Life-altering innovations will include worldwide high-speed communications, robotics, artificial intelligence, voice and thought brain-to-computer interfaces, and quantum computing. Medical innovations will see stunning progress in lowering mortality rates and prevention and early detection of diseases of aging. Democratizing products and services is one of technology's highest and best uses because it lifts the living standards of all, particularly those at subsistence levels.

Shrinking distances

The next decade will see rapid advancement in both physical and electronic travel.

Innovations in physical travel will make the world smaller and accessible to more people, even as options to interact electronically will make more physical travel optional. The COVID-19 pandemic accelerated these changes.

THE NEED FOR DEMOCRATIZED WISDOM

The wisdom deficit

Wisdom advances the development and maturation of individuals, societies, and nations. The more complex and sophisticated people and communities become, the greater the demand for wisdom.

A new and better educational mission

Change forces are disrupting education, just like all institutions. College and K–12 enrollments are declining with the birth rate. Many employers are dropping college degree requirements. Public policy polarization and uneven performance of public schools have increased interest in public school alternatives. Expanding high-speed internet access is making knowledge and information a commodity. People will go through many more jobs and careers over a longer life, making life-long learning a twenty-first-century requirement. Education's content and delivery model need upgrading to address these and other New Reality demands. A new educational mission and delivery model will be developed within a decade to remake education into on-demand, personalized, lifelong learning, hopefully with wisdom and better decisions as its mission.

Societal complexity and decision-making speed

The societies emerging over the next few decades will be far more complex and change more rapidly than anything humanity has yet experienced. The best preparation is adaptability and being well grounded, calm, and fearless. Those qualities define every wise person I have known. Wise people come from all walks of life, careers, spiritual beliefs, and economic strata.

Cultivating wisdom in young people to make better decisions is the most realistic, efficient way to build a better twenty-first-century society.

Human relationships with automation

Our relationships with automation will be more complex and require more attention as automation becomes more humanlike, seamless with our lives, and prevalent to offset declining populations. Each person must define the boundaries, controls, and timing of their integration with automation and their relationship with it.

More responsible individuals

In a democratizing world, fewer decisions trickle down through the traditional top-down hierarchy, leaving more decisions to individuals from the bottom up. One example occurred during the COVID-19 pandemic when the general public lost confidence in government and public health institutions. Instead of waiting to be told the pandemic was over, they decided that for themselves. Expect more of this bottom-up change to occur in the future.

BECOMING WISE IN A FOOLISH TIME

Widespread achievement of potential and wisdom may seem impossible, given the pervasiveness of foolish behavior and poor judgment today. Necessity is the mother of invention, however. Norms change when the need becomes urgent, and the old ways become unacceptable to the majority. Hard times precede better times. Some of the worst times in Western history led to its most notable heights. In every generation, some people choose to live to their potential even though it is out of step with those around them and, at times, even dangerous. Many of the people we honor today as visionaries were once considered heretics or insane; they were wise beyond the time in which they lived.

In this book, you will read stories of people called to reach their potential and live wisely. Some described it as "an inner summons." Others experienced an awakening event. Some had an "unclear ambition" for something beyond the norm. All responded, rejected the ordinary, and refused to follow the herd, even as their contemporaries conformed. These Potentialist pacesetters bequeath a wealth of insight about the benefits and practices of pursuing one's potential.

HOW CAN YOU REACH YOUR POTENTIAL AND BECOME WISE?

AN INNATE GROWTH MECHANISM IS WAITING TO ASSIST YOU

Realizing your potential and becoming wise is as natural as breathing. It is similar to optimizing your physical body's performance. Humans are the only species that must choose to cooperate with their innate growth mechanism.

Your potential is your best self that I call the "I" in "Me" or "INME." It resides just below the surface of your conscious awareness, develops over a lifetime, and its growth can be accelerated. You likely already experience the INME when you have an insight or say something wise and think, *I didn't know I had that in me.* Later chapters explain how to call on the INME when and where you choose. In time, it becomes established and is always conscious. You will not struggle with disappointing yourself or others with behaviors or attitudes that surprise you in unpleasant ways. As you will learn, they are not your own but part of an innate defense system.

As the INME develops, you become incrementally wiser and more prepared to address difficult decisions. This is because only your INME can be wise, for reasons explained in the coming chapters.

POTENTIAL AND WISDOM ARE LESS COMPLICATED TO ACHIEVE THAN YOU MAY THINK

I began pursuing my potential when I was twenty years old. It became a passion that shaped me, my leadership style, my entrepreneurial strategy, and my devotion to family. During the half century since, most people have continued to believe that wisdom is only for special people because it is so complex. That simply isn't true. You already know and do much of what is needed to achieve your potential and wisdom. You do it unconsciously, sporadically, or for a different purpose. Using these known skills consciously, purposely, and routinely enables you to pursue your potential and wisdom easily and efficiently.

Widespread appreciation of this fact and adoption of the needed practices is how potential and wisdom can be democratized. The following are a few of the capabilities, skills, and practices you and almost everyone already know that can be redirected to potential and wisdom's pursuit. Many more are described in the following chapters.

Vows and pledges from the heart

We learn the Pledge of Allegiance to the United States as children, and most of us take it seriously. I once watched several hundred new citizens make their first pledge, most with tears streaming down their faces. We make vows and pledges to our lifelong friends, spoken and unspoken, to have their back. We make marriage vows with serious intention.

A simple, heartfelt daily commitment to yourself launches the pursuit of potential and wisdom: "I will do my best, to be my best, and leave the world and the people I meet a little better than I found them."

This simple commitment aligns you with your innate growth mechanism. It isn't a slogan, a club, or a feel-good motto. It is a personal, private, heartfelt, all-in promise to yourself that defines every aspect of your life and becomes its ultimate measure of success.

Visualize to realize

The same process everyone routinely uses hundreds of times in life to learn and practice new skills at work, in sports, and in other pursuits is an effective, scientifically proven method to reach your potential and become wise. A century ago, psychologist Carl Jung called it Active Imagination, borrowing many practices from theologian and philosopher Saint Ignatius five hundred years earlier. All that is necessary is to consciously, deliberately, and regularly use this known skill to have a powerful method for realizing your potential and wisdom.

Reality beats illusions every time

You will soon learn that much of the difficulty and heartache we experience comes from seeing life through a lens distorted by our Ego. You probably find that hard to believe at this point, but stay with me, and you will see exactly what I mean. Life looks better and makes much more sense when you see through a clear lens. Anyone with common sense can have a wise person's perspective when you learn to detect and avoid Ego distortions and illusions.

The wind beneath your wings

Pursuing your potential and wisdom is a wholly personal and private journey. But that doesn't mean you do it alone. Almost no one does. First, you have me and this book! You also undoubtedly have wise people around you who will become mentors and guides. With your encouragement, some can become your very own versions of Socrates and Aristotle, posing challenging questions that develop your potential and wisdom. If no one currently comes to mind, you can find them once you learn in this book precisely what wisdom is and how to spot it in others. I tapped hundreds of such people in my lifetime. A handful become co-creators; you shape them, and they shape you. A level of intimacy develops that lasts a lifetime.

Wisdom—assist artificial intelligence

Technology exists and will rapidly advance to connect you to the wisdom of the ages and that of people you trust. It will be personalized to your specific interests and situation to accelerate your capacity for wisdom. It will also be available to you on demand when faced with complex decisions requiring wisdom. It cannot and should not make decisions for you but serve as a ready adviser instead. The technology will become a societal norm over time and accelerate the democratization of wisdom.

A universal ethos

This book is written for you in the most understandable style I could craft with the goal of constructing the largest possible "welcome" tent. I used secular rather than religious terminology when discussing psychological evolution, the Soul as a function, and creation, hoping this book will feel comfortable to all.

When necessary, new terms were used, such as INME instead of "self." "Intimacy" is used interchangeably with "kinship" (meaning from one seed) when referencing the connection between all things in creation. The term "psychological body" is used to compare and contrast our physical and psychological existence.

Experiences and observations play a large role in this book and its underlying processes and practices. For over half a century, I observed, studied, interviewed, or read accounts of the wisdom of hundreds of wise people, from philosophers to office cleaners, gleaning what makes some people wise and other people fools. That manifests in this book as detailed but comprehensible and observable profiles. They are used to assist you in recognizing, relating to, and emulating wisdom and the psychological functions of Ego, Shadow, Soul, Conscience, and INME at work in people. "Wisdom Stories" close the first ten chapters. They are about real people who faced tough situations, demonstrating one or more qualities of wisdom.

Psychological theories and practices of self-actualization or individuation are in many ways similar to the pursuit of potential and wisdom. This book synthesizes and simplifies them in laypeople's terms emphasizing the observable result as potential and wisdom. The goal is to shape a universally acceptable ethos in the spirit of psychology's founders and modern practitioners.

A CIRCUMAMBULATION INSTEAD OF A LINEAR PATH

Your pursuit of potential and wisdom will be dynamic and directional but not linear. Unfortunately, books must be read front to back, but your growth will be wavelike and organic rather than serially like chapters in a book. Dr. Carl Jung, in his autobiography, wrote, "I began to understand that the goal of psychic development is the self [INME]. There is no linear evolution; there is only the circumambulation of the self."[1] It is a new way to learn, but it will blow your mind once you get it working for you. You become healthily addicted to your growth, which is sometimes dizzying and other times slower. You feel joy and intimacy like you never believed possible. Above all, trust the process and keep moving forward.

A USER'S MANUAL FOR THE REAL, BEST YOU

Early twentieth-century people could not have imagined that we would have humankind's collected knowledge at our fingertips, global cellular phones, video-teleconferencing, internet service, travel by jet airplane anywhere in the world within hours, and widespread use of artificial intelligence. It may be challenging to visualize most people operating at their potential, living wisely and well through better decisions. **Human potential has always been underestimated. Humanity, and especially its leaders, have yet to grasp this timeless truth.**

Everyone has a stake in this societal step change to democratize wisdom. Young people must grow up faster and wiser to accept the responsibilities of a democratizing society. Midlife and older adults must be willing to be their Socrates or Aristotle. Millions sacrificed their lives for us to live as we do today. We owe it to them to do our best, to be our best.

Your first act as an emerging twenty-first-century pacesetter is to visualize yourself and your life as they can be instead of what they are today. So prepare yourself. Look in the mirror. Do you sense the greater you? An unimaginable future awaits. **You are never too young or old to learn, grow, and become wiser.** Your future is calling.

WISDOM STORY:
WHO MADE THE GREATEST GENERATION?

This story of wisdom is from Wade Dyke in his voice. Wade's calm, quiet demeanor is an outer manifestation of a humble, brilliant, remarkable man. His story is an insightful reminder of a theme throughout this book that how we live and influence others has effects far beyond anything we can imagine. Wade's story is particularly astute for these times of great change and everyone's responsibility through this transition.

Americans who grew up during the Great Depression and lived through World War II are commonly referred to as the "Greatest Generation." They are recognized for their wartime courage, commitment to civil rights progress, and resilience shaped by economic hardship. We rarely consider the wisdom of the generation that raised them. My grandparents were of that generation, including my grandfather Glenn Piper.

Glenn was born in 1895 and grew up on a northern Illinois farm, the oldest of five siblings, three boys and two girls. One sister died young. Glenn joined the army at age twenty-two during World War I to assist France, a steadfast US ally since independence from Britain. For religious reasons, Glenn didn't want to shoot anybody but volunteered to be a runner instead. Runners traversed trenches carrying messages and battle instructions from command centers to fighting units. The job was often more dangerous than being a combat soldier. Glenn was gassed and wounded, but he survived.

He loved farming and returned to it after recovering from war injuries but lost his farm in the Depression. Glenn became the first in his family to move into town to make a living, working thirty-five years for the post office. It was his plan B, though he joked that being a runner prepared him to deliver mail. He married and had three daughters, each uniquely contributing to the evolving economy. Glenn had only an eighth-grade education but taught himself the stock market. The PE ratio was his preferred statistic. He read avidly and saved quotes from literature. A framed Tennyson poem, "Crossing the Bar," held an honored place on his bedroom dresser.

The courage, compassion, and resilience of the Greatest Generation came from somewhere, perhaps from the greater generation of their parents who put food on the table in tough times and guided them through their youth. American history reveals many such generations, each making extraordinary contributions and passing the qualities to their children. When we witness wisdom in others or generations, we should remember that someone inspired and shaped them. We all have that opportunity and duty within families and outside by the examples we set and by mentoring and coaching younger people. We can hope that a free citizenry, motivated by opportunity, faith, and curiosity, retains such character for generations to come.

CHAPTER 2

Life's Unavoidable Questions

"Those who cannot change their
minds cannot change anything."

—GEORGE BERNARD SHAW

HAVE YOU EVER THOUGHT OR SAID SOMETHING that left you feeling bad or unsettled for saying it? Or did you agree with someone else's statement or fail to object to something said that didn't feel quite right? Perhaps it was a derogatory statement about life, people in general, a segment of society, or a dystopian outlook on the future. Reflecting on it, you weren't sure about the accuracy of what was said or where that attitude or opinion came from. Many assumptions, attitudes, and convictions we hold are not our own. They are distortions or illusions of reality coming from society's collective Ego, our own Ego, or others' opinions that we accept unchallenged.

Sometimes, these distortions cause severe distress and disruption in our lives. Psychologists call them "cognitive distortions." Many have been specifically enumerated and defined. Other less-defined types are sometimes called Ego "illusions" or "fictions." Much of a therapist's work in cognitive behavioral therapy is helping patients overcome distortions that cause pain and suffering. Even when not severe enough for therapy, unaddressed, distorted, or limited perspectives about life's foundational issues and questions cause problems in careers, relationships, and society.

Fortunately, a practical, proven method is available to build a strong foundation of accurate perspectives. Life's foundational issues and attitudes can be posed as questions. By consciously confronting, critically examining, and answering them, perspectives become more well-grounded and your own. Every question is about sharpening your perspective of your worldview and yourself. Many ancient cultures used this approach. Greeks, Romans, and others used a similar teaching method with their youth. Today, classical educators utilize the same process, examining many questions through classical literature. The questions may be referred to as existential, philosophical, or unavoidable questions. *Unavoidable* means that failing to address the question consciously and deliberately results in unconsciously accepting someone else's opinions as your own.

Life-changing benefits are derived from addressing unavoidable questions. To be wise, we must seek truth and reality in the outer world and our inner world instead of living in an illusory world created by the Ego or the collective Ego in society. How can we know who we are, be ourselves, or become our best if life's defining questions are never addressed? You risk awakening one day, wondering how you came to be who you are, living a life you did not choose. Answering foundational questions prepares you for decisions where moral clarity is essential. You grasp reality as older, wiser people usually do, equipping you to make wise decisions earlier in life. Discussing unavoidable questions with a wise mentor, friend, teacher, or therapist can be helpful.

This chapter examines some of life's most important unavoidable questions and illustrates the depth of inquiry and analysis every question deserves. Use them as models to identify and answer other unavoidable questions, some of which are raised throughout this book. The unavoidable questions are set out in green italic font.

DOES MY LIFE AND HOW I LIVE IT MATTER?

This question explores the importance of every life, beginning with your own, and how failing to value life leads to alienation and despair.

BILL'S STORY

Bill has had a long, difficult week. He arrives home after another stressful day and a brutal commute. His years of hard work paid off until his boss recently retired. His new boss doesn't appreciate Bill's work, years of service, or dedication. Today, they argued bitterly. Bill loves his job but may be forced to quit or wait to be fired. He is bitter and disgusted by the unfairness of it all.

Exhausted and exasperated, he pulls into the driveway of his heavily mortgaged home, craving a shower and a beer. As he opens the front door, the family dog runs past him and down the street. He doesn't feel like chasing the stupid dog but knows he will catch hell from his wife and kids if he doesn't. After wasting an hour retrieving the mutt, he grabs a beer just as his phone pings a text from his wife. "I'm on a conference call until 7:30. Can you make dinner tonight?" *Great, just what I need after the day I've had,* Bill thinks. He grabs the beer, pulls pizzas from the freezer, and pops them in the oven before heading upstairs for a shower.

Bill comes downstairs refreshed and is met by the smell of burned pizza. His wife is in the kitchen in a fury. "I asked you to do one thing, and you burn frozen pizzas?" In seconds, they're in a full-fledged fight. Bill storms out with a ruined appetite. He slams the door to his home office, pours a scotch, and stares into the darkness. *Why do I bother? All these years of hard work for my company and family, and for what? Does my life and how I live it even matter?*

Everyone has moments like Bill's when we question our value and life's meaning. Those times cause us to lose perspective and become alienated.

If severe or frequent, resentment, stress, victimhood, and bitterness can build. Intimacy with loved ones can become strained or permanently damaged. Physical and mental consequences can last a lifetime.

It doesn't have to be this way. A characteristic of the wise is a firm belief in the value of every person's life and their opportunity to contribute. Understanding the gift of life and your irreplaceable value begins with an unshakable answer to the question: *Does my life and how I live it matter?*

PEOPLE UNDER STRESS CAN LOSE PERSPECTIVE

Occasionally, any of us can feel beaten down, injured, or defeated. Usually, we brush away doubts about life's meaning and value and deny the unpleasant feelings. Leaving the question of your life's value unresolved makes you more vulnerable the next time a lousy day arrives. Indecision causes you to waver, use poor judgment, give up on life, and become chronically alienated. An accurate perspective of life and our value places us on solid ground, prepared to face and work through adversity.

THE GIFT OF LIFE AND THE MACROCOSM-MICROCOSM

Answering whether your life and how you live it matters begins with the miracle of life itself—that you and all other humans exist at all. Multiple evolutionary dead ends were possible in the millions of years preceding the emergence of the human species. An ecological disaster like dinosaur extinction could have ended humanity's march at any time. Yet . . . here you are.

Furthermore, you don't simply exist; you know that you exist. We humans alone, of all species, can ponder the creative force, nature, natural law, or God (whichever you find most comfortable) that created us and the universe. We can use science, reasoning, observation, and spiritual contemplation to examine our existence, find meaning, and discover our place in life's grandeur. Science and mathematics argue infinity is full of undiscovered life, and we can hope for it. But for now, it's just us.

It is astonishing and reassuring that whatever created the universe finds little old us as necessary as new star systems or inventing another species. A beautifully written article, "This Is Why You Matter: The Theory of Macrocosm & Microcosm," attempts to explain and reconcile scientific and mathematical theories of the universe. Author Alex C. Wilson says:

> Each individual is a reflection of the entire universe—as in the macrocosm, so in the microcosm. Essentially, what that means is this: whatever is happening in the universe at large is also happening within each of us. If you've ever felt even a little bit insignificant as one of the billions of people on this planet, this theory proves that the opposite is true; you are far more significant than you could ever imagine.[1]

THE MACROCOSM-MICROCOSM IN MATH, SCIENCE, AND TECHNOLOGY

Like the work of Einstein and others before him, Quantum theory challenges the foundations of the universe and how we view life. The Ego frames life as binary—on/off, either/or, black/white. Quantum reveals that something can be neither, either, or both simultaneously. In time, quantum applications will change our lives dramatically. Quantum computers, thousands of times more potent than today's, will solve problems current technology could never solve. They might include space travel, near-perfect weather predictions, and human body simulations to prevent disease and extend life. Equally important, quantum could alter how we define life and our place in it. The macrocosm-microcosm may be explained, and other explanations may prove how we can be both individual and mortal yet infinite. Perhaps life and death as we know them will be redefined.

We do not have to wait for quantum physics to understand practical applications of the macrocosm-microcosm and both. With its individually addressable users and devices, the internet demonstrates how something massive and universal can know and personalize its relationship to each of us, empowering both it and us. Humanity is freeing itself from historic limitations of time, distance, language, culture, education, and economic status. Internet cloud-based applications cater to individual differences, ultimately enabling a bespoke, personalized world. But if we all stopped using the internet tomorrow, it would cease to exist for practical purposes. In a small way, the relationship between the internet and each of us demonstrates the macrocosm/microcosm theory that without one, there is no other.

The personalized internet-to-user relationship is a human imitation of creation. **Each of us relates to the greater (creation) through the psychological function we call the Soul,** like an individual IP address. The internet/cloud is the puny (by comparison) proxy for creation and infinity. The internet personalizes and communicates with each user, similar to how the Soul function personalizes creation to each of us.

You can observe the macrocosm-microcosm in your daily life, thoughts, nature, the universe, science, international affairs, and relationships. You are an integral part of the larger fabric, a necessary piece of an expansive tapestry that extends far beyond the confines of your physical and material reality. You will see the same interrelatedness at work all around you. That perspective can help frame decisions differently. Considering that, reconsider the question: *Does my life and how I live it matter?*

THE ISOLATION ILLUSION AND THE CONNECTEDNESS REALITY

You are a member of the human community of roughly eight billion people. Billions went before us and will follow us. Yet, walking a few blocks or miles brings you in contact with kind people. You share an exquisitely beautiful, diverse, natural world with all living things. Gazing into the sky connects you to the infinite. Isolation diminishes us, making us feel irrelevant in a vast horde. It is an Ego distortion explained above. The macrocosm-microcosm is the antidote, showing us that we are never alone. We are a perfect creation, inseparable from the macrocosm of all.

Consider the natural wonders found in water, beaches, and snow as a demonstration of microcosm-macrocosm. Together, they form an extensive system of life-essential water. Each body of water, beach, and snowbank is formed from vast quantities of unique individual water drops, grains of sand, and snowflakes, just like unique individuals form humanity. Seventy-one percent of the earth's surface is water in oceans, seas, lakes, and rivers. Beaches are natural borders between water and food-producing earth spanning fifteen times the earth's circumference, over 372,000 miles. The water from the snowpack and seasonal snow covers 46,000,000 square kilometers, about a third of the earth's land area.[2] The vastness of these natural phenomena and their life-enabling interconnection is awe-inspiring. Earth is the only planet known so far to have such a system.

You are essential to humanity and infinity, like each drop of water, grain of sand, and snowflake. **We choose what we make of being an act of creation by constructive creativity in how we live and treat others, and thus alter infinity.**

ARE WE HUMANS EVOLVING TO EXTINCTION OR OUR POTENTIAL?

You have likely witnessed discussions where a case is made for a downward spiral of life and humanity on the verge of extinction. This question explores the human fascination with extinction and how it supports a self-perpetuating and self-sabotaging illusion of powerlessness.

THE EXTINCTION MINDSET PHENOMENON

People have been attracted to predestined, imminent extinction since early recorded history and perhaps since humanity's beginning. Extinction has remained a recurring theme for millennia.

Today, doomsday dominates headlines and conversations and creates an undercurrent of anxiety. This is despite all past extinction predictions being wrong and overwhelming evidence of human progress. People are better educated, more civilized, and perhaps more intelligent than ever. Prospects are bright, though never guaranteed. Shouldn't we be independent and critical thinkers focusing on something other than a primal, generalized, and unactionable fear for most of us? What's going on? Are extinction threats worse than in the past? Is life more dangerous? Could life be safer but seem more hazardous because alarmism and exaggeration are effective ways to sell us on a cause, product, or personality through a twenty-four seven worldwide mass communications system?

These are essential questions about forces that act upon us, but it's equally important to consider why extinction predictions remain seductive. What are the consequences of an extinction mindset to individuals and society?

CONSEQUENCES OF AN EXTINCTION MINDSET

Pessimism closes our eyes to opportunity, and decisions based on fear are almost certainly unwise. Extinction beliefs are prone to becoming fanatical. Over the centuries, people died in suicide cults that sought to cheat the predicted cataclysmic event by dying first, odd reasoning at best, if not downright nutty. Extinction fanaticism is not a historical relic; it is alive today. It can be found in causes that attempt to justify murder and terrorism, in extreme advocates who violently oppose any theory other than their own, or those optimistic for the future. It can be found in conspiracy theories and extreme survivalists.

Younger people are particularly susceptible to extinction exaggeration and drama. It has an indisputable though unmeasured influence on depression, violence, and suicide, the second leading cause of death among young people aged twenty-five to thirty-four.[3] Fearmongering certainly can't help those struggling to find a reason to live. It steals joy and happiness from young people who should be optimistically pursuing their potential, intent to shape their lives and contribute to the world at large.

Tragically, some young people have been convinced that they should not bring children into "this awful world." This twisted logic disregards factual realities and guarantees the extinction of humanity if enough young people take the bait; the population is already declining precipitously. These young people will miss the joy and meaning derived from children and grandchildren. I've met far too many people late in life who regret the decision not to have children. It may suit some, but it should be carefully and wisely considered, not fueled by an extinction myth.

Predators learned long ago to use fear to manipulate, and it remains a powerful and pervasive strategy for making a sale or gaining people's allegiance. Few fears are as potent as the fear of death and extinction, which makes it a powerful tool for those intent on amassing wealth or power at any cost. Today's predators include cause extremists, the leadership of countries hostile to the United States, Western civilization, and democracy, and some irresponsible politicians, media outlets, educators, and product marketers. **An extinction mindset is corrosive to individual and societal growth, human potential, and the enjoyment of life.** It distracts us from urgent, actionable threats and predictable but undefined threats called Black Swans. Known risks fuel innovation to address them. Unknown threats are more dangerous because we are unprepared for a response, as demonstrated by 9/11, the 2008 financial meltdown, and the COVID-19 pandemic.

Perhaps most dangerous, an extinction mindset stunts the perspective required for wisdom and wise decisions. Perspective requires an informed, unbiased examination of ourselves and humanity's past, present, and future.

DISTORTED PERSPECTIVE—IGNORING PAST PROGRESS

A history professor of mine once said, "Human existence until recently meant most children died before twelve, few lived past fifty, and life between was dominated by hunger, disease, and violence. Today, we feel cheated by anything less than peace, full bellies, and an ever-increasing lifespan past sixty-five." He was right; expectations today are highly elevated. Humanity took a long time to emerge from life as hunter-gatherers and subsistence farmers. But our ancestors endured, innovated, evolved, and gave us a chance for a better life.

Fast-forward to Europe's Middle Ages from the fifth through the fourteenth centuries, a period glorified in literature and the arts by castles, court ladies, and knighthood. Civilization advanced slowly during those nine hundred years with starts, stops, and significant setbacks that included some of the most brutal times in recorded history. Later, the first of three waves of the bubonic (Black Death) plague killed an estimated twenty-five million people, almost a third of the population, in only four years, 1347 through 1351. The plague endured for centuries, killing millions more. The people of the Late Middle Ages had reason to believe that humanity was in its final act. Unsurprisingly, apocalyptic thinking prospered.

But people did not give up. Horrible times gave way to reflection, hope, and action, ushering in the Renaissance and Age of Enlightenment. Pacesetters looked to past civilizations to learn how to make life better. Discoveries in science, engineering, medicine, innovative philosophy, and religion called people to their higher selves and better times. Ships connected continents, opened trade, and shrank the world. Art, creativity, potential, hope, and innovation replaced disease and death as life's dominant themes. People endured, innovated, and evolved, giving us a chance for a better life. In the mid-1890s, the cause of the Black Death was finally discovered. Rapid urbanization, overcrowding, and lack of sanitation had led to an infestation of rats carrying fleas with plague.[4]

Fast-forward again to the considerably more sophisticated twentieth century. Author and futurist Peter Diamandis, MD, is a twenty-first-century pacesetter who writes extensively about negativity and pessimism distorting factual abundance. In one blog post, he explains the dark times people endured from 1899 to 2023, with 265,524,000 deaths from war, famine, the Great Depression, and pandemics.[5] He reminds us we live in a time of relative safety and prosperity with a bright future. The people of the twentieth and early twenty-first centuries endured, innovated, evolved, and gave us a chance for a better life.

Given humanity's history of progress, why do so many lose sight of the gift of life and the price others paid so we can live free and well?

DISTORTED PERSPECTIVE—
ACTIONABLE AND LESS LIKELY THREATS

All living things age and die. So far, science can't make us immortal. But we aren't helpless, either. We can and should do everything possible to live a long life by lowering the risks of premature death, but many people don't.

Below are the top ten causes of death in 2021, accounting for almost 75 percent of US deaths, with a few 2022 and 2023 updates where data was available. The list is relatively unchanged from a decade ago, except for COVID-19.

- Heart disease (695,547)
- Cancer (605,213)
- COVID-19 (2021: 416,893; 2022: 244,000)
- Accidents (224,935, including 46,980 auto accidents[6])
- Stroke (162,890)
- Chronic lower respiratory diseases (142,342)
- Alzheimer's disease (119,399)
- Diabetes (103,294)
- Chronic liver disease and cirrhosis (56,585)
- Kidney disease (54,358)[7]

Unlike our ancestors, we know how to improve our odds. Follow a healthy lifestyle of diet, exercise, sleep, and quiet time. Develop strong, intimate relationships. Avoid reckless behavior proven to increase risks of death or injury, such as addictions to smoking, alcohol, and drugs, speeding, reckless and drunk driving, and home accidents, especially accidental poisoning, and falls.

You and your loved ones will almost certainly die from a cause above or one of the lesser causes, not an extinction event. Some people disregard preventable, actionable risks but become absorbed by extinction threats. A friend said he watched a fellow in his thirties chain-smoke during an hours-long environmental protest. The club of people powerful enough to reduce extinction risks is tiny, and you and I aren't members. Isn't reducing known, actionable risks wiser than devoting energy to unactionable, less likely extinction risks?

DISTORTED PERSPECTIVE—
A WORSE OR BETTER FUTURE?

Apocalyptic mindsets are built on falsehoods, alarmism, and analytical failures. They minimize or demonize humanity's progress but accept mass media negativity as reality. They mistakenly rely on the Ego's fear-fueled linear thinking that present-day threats will only worsen, ignoring historical examples of human ingenuity addressing known risks.

Apocalyptic mindsets equate optimism with denial of extinction risks. Pragmatic optimism accepts extinction threats and the causes of death above and weighs the preponderance of evidence and experience to determine the wisest course. **Pragmatic optimism suggests the better argument is that humanity is evolving toward its potential rather than its destruction.** We might blow it, but history and momentum say we will endure, innovate, evolve, and create a better life.

WHAT MAKES US RECEPTIVE AND
VULNERABLE TO EXTINCTION THREATS?

Some possible reasons include the following:

1. Our Egos crave certainty and control, but they are illusions. Reality is uncertain, with very little that is controllable. The constancy of change perpetuates uncertainty. As the pace of change accelerates, the perceived loss of control and certainty causes many people to become fearful and anxious. They invent or adopt fictional scenarios that appear to offer certainty and control, such as conspiracy theories, survival myths, and extinction.

2. Twenty-four-hour internet and cellular access exposes us to fear-based sales and manipulation.

3. Widening social division unnerves us. We'll explore this issue more in the next chapter.

4. Our psychology hasn't adapted to accelerating change. There is so much change happening so fast that we cannot accurately assess what is positive or negative. Fear of all change increases to the point of extinction physically or in terms of who we can and will become.

5. For hundreds of years, religious worship connected most people to their Soul and Conscience functions, compensating for fear. As religious practices have declined, no practices have replaced the lost connection.

6. Young people grow up more slowly as life expectancy expands. They are more susceptible to alarmist manipulation and focus on external factors that they cannot control rather than focusing on their reactions to external factors that they can control.[8]

An essential New Reality adaptation is acceptance that there have always been and will always be extinction threats. Our reaction to those threats is what we can influence. Living to your potential by making wise decisions is your most effective response. The wiser you and others are, the better you will understand, accurately assess, and respond to such risks.

QUESTION 3:

ARE PEOPLE COMPOSED OF MOSTLY INFERIOR QUALITIES, SUCH AS EVIL AND LAZINESS, OR ADMIRABLE QUALITIES, SUCH AS GOODNESS AND INDUSTRIOUSNESS? ARE PEOPLE GETTING WORSE OR BETTER?

These questions explore perspectives and attitudes toward our fellow humans, the origins of those perspectives, and how to adjust them.

We uncomfortably witness people deceiving, manipulating, cheating, or stealing from others or collectively from all of us through illegal or immoral actions that increase prices, insurance rates, or taxes. Even some who are wealthy, socially prominent, or professionals at the top of their game unethically disadvantage others. We see people wantonly kill others when the human race has come so far. Have you ever considered how much more enjoyable life could be and how much more disposable income you would have if predatory behavior radically declined?

Many people conclude that humans are fundamentally rotten. You've undoubtedly heard this in conversations. Probably all of us have felt it momentarily. Like attitudes toward extinction, some people seem to enjoy holding and proselytizing negative opinions about human nature, even though it is depressing and unactionable.

CAN GENERALIZATIONS ABOUT PEOPLE EVER BE VALID?

Generalizations are imprecise approximations at best. A statement that begins with "all people" must be suspect beyond a few exceptions. *All people* being subject to the laws of physics is true. *All people* being inherently evil exaggerates to the extreme. How many of the almost eight billion people on Earth would you need to know well to state that all think or behave a certain way?

Sit in any cocktail bar or restaurant and listen to nearby conversations. Gender, ethnicity, appearance, and other biases are prevalent among decent, sophisticated people who fall into automatic thinking, which is something like muscle memory. Such statements may have harmless or humorous intent, but the mind's efficiency reuses generalizations unless trained otherwise. Everyone can fall into automatic thinking, even if consciously unbiased. Automatic thinking pops up when we are emotional or anxious.

However, generalizations are sometimes helpful in conceptualizing possible meanings behind enormous numbers with extreme complexity. Communication and creativity would be stifled if every discussion required a statistically supported, peer-reviewed citation. However, overused generalizations lead to unwise, untested, biased mindsets and decisions. There is a middle ground we can all follow and challenge others to do so as well.

Using or accepting absolutes about humanity or large populations is a bad habit that perpetuates an illusion of uniformity that does not exist. It is an easy habit to break and to influence others to break as well. Be as specific as possible to increase accuracy and credibility. Check for data from a credible source before generalizing, which is made easier by the internet and smartphones. Substitute absolutes like all, none, and never in favor of terms purposely less precise, such as most, some, a few, and sometimes. Qualify generalizations using the bell curve or standard distribution that is based on the natural law of diversity, as explained in the next chapter. Most people recognize the bell curve from high school, college, or careers. Humanity has an enormous range of physical and personality characteristics. They are mainly uncatalogued and undefined today, but they will be complete and organized for study someday.

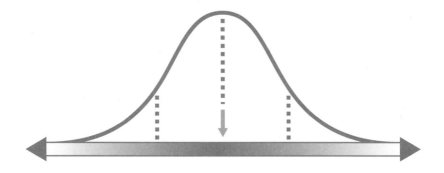

Consider how an answer to the question *"Are people inherently bad or good, getting better or worse?"* becomes more manageable and clearer using the bell curve. It remains an approximation but is closer to reality. A small percentage of people seem to have no sense of right or wrong, presenting a danger to others; their opposites are so virtuous as to seem saintly. The rest of humanity falls in between. Imagine the majority divided by a line: half are predisposed by degrees to virtue, and half are predisposed by degrees to harming others. The bell curve is a practical and more accurate perspective of human diversity.

Imagine you are at a social or business gathering, and someone uses a negative generalization like "Today's kids are all lazy." You can disrupt the mind's automatic thinking with the bell curve. You can politely inter-ject, "Whoa, do you have any data to back that up? Or is it an anecdotal example or your opinion? Isn't it more likely that young people fall in a bell curve in work ethic, like everything else?" This is one way to elevate wisdom in discussions rather than encouraging bias and intellectual neg-ligence. In future chapters, you will learn more ways to disrupt automatic thinking, such as changing your state of mind or mood.

ARE PEOPLE GETTING BETTER OR WORSE?

Crime statistics, data on social norms, and experiences suggest that people are more civil and moral today than ever. Only a few centuries ago, bearbaiting with dogs was a sport, and people were routinely burned at the stake for differing religious beliefs. Only a century ago, vigilante executions without trial were commonplace in some parts of the world, including the United States, with children invited to attend. In World War II, people supporting the Allies and Axis celebrated the deliberate mass bombing of civilian populations. Today, accidental noncombatant deaths cause an immediate outcry. The punishment for many crimes for millennia was execution, often brutal and public. Today, capital punishment is rare, mainly limited to premeditated murder, and prohibited in much of the world. And on and on.

We encounter predators more frequently today because the internet and cellular technology are an easy route into our lives that didn't exist before. Also, people rationalize predatory technical behavior because they don't witness the harm they cause, and the risk and consequences of being caught are lower than those of physical crime. So we experience humanity's worst firsthand more frequently than in the past. Hopefully, cybercrime protection will catch up.

Technology is rapidly enabling worldwide multilingual communications, soon with no computer skills required. That will enable universal, affordable education and economic opportunity, reduce subsistence living, and expand human potential. How will human civility and predatory behavior change as a result?

How often are perspectives of humanity an unconscious self-assessment projected onto others?
Every unavoidable question requires examination of both the outer world of the macrocosm and our inner world of the microcosm to achieve an accurate perspective. A time-tested observation of

human nature is that someone excessively professing a quality like honesty or truthfulness is less likely to possess or practice the quality. Excessive jealousy and possessiveness can signal a lack of confidence to be faithful. Does it hold that the person with the dimmest views of humanity is the most unsure of their character and judges others as harshly as they judge themselves? Emphatic, chronic use of absolute terms increases the likelihood of a person projecting an internal conflict. The Ego makes you feel better about being rotten by convincing you that all of humanity is equally rotten.

Each of us has talents and areas that need development. We make mistakes and have said and done things we later regret. We have Souls and Consciences along with Egos and Shadows. We can be kind and generous but insensitive. Are we the sum of wise decisions and acts or those that are unwise? Our best moments or worst? If human nature is inherently evil, so are we, our loved ones, friends, neighbors, and even newborns. Harsh judgments are the work of a defensive Ego and Shadow instead of our best self, the INME.

It's healthy to hope, wish, and work for human potential, starting with our own. I've devoted my life and this book series to that goal instead of judging and bemoaning human deficiencies. Heroes like Abraham Lincoln, Benjamin Franklin, Eleanor Roosevelt, Mahatma Gandhi, and Mother Teresa should be celebrated for their contributions without expecting perfection. Their acts of kindness and courage shaped history more than tyrants and villains and contributed to a collective aspiration for who we become.

So do the acts of billions of people who unselfishly help others daily, like the teenager who aids a ten-year-old who fell off a bike, and the patient driver who forgives another's driving mistake by remembering their own. Every day around you, the best of humanity is displayed when our eyes and ears are open. Humanity is like each of us—complex, contradictory, seeking its potential, and, on the whole, consistently getting better.

DETECTING AND ANSWERING UNAVOIDABLE QUESTIONS ON YOUR OWN

This chapter has introduced you to three unavoidable or existential questions. The remaining chapters will introduce you to others, or you can visit PotentialistFuture.com for more discussions of unavoidable questions.

You can also develop your own unavoidable questions by noticing topics and situations where your footing is unsure or that stimulate your curiosity. Maybe a broken relationship causes you to wonder, *What is the nature of forgiveness, and how does one know when and how to forgive?* Perhaps self-reflection causes you to ask, *How does one identify a meaningful career?* Even a world event or trend might raise a question: *Does the technology shrinking the world advance greater commonality, diversity, or both?*

The process relies on curiosity, depth of inquiry, and asking and answering questions. The goal is a broader, deeper, wider perspective based on reality and timeless truth. Your answers define you and your life's course. Engaging teenagers and young adults in discussions of life's unavoidable questions is incredibly beneficial. It takes patience, but they will appreciate your intentions and your wisdom at the right time in the future.

WISDOM STORY:
SEEING AND ACCEPTING REALITY

Aaron was a cable television pioneer who later founded several highly successful companies in diverse industries, including one whose market value today is $1.5 billion. He was a visionary who saw rural community cable television as a terrific idea with perfect timing, but that didn't make success come easy.

Few people understood the new Community Antenna Television (CATV) technology or its service challenges, especially the city officials with regulatory authority to select a single cable provider. Cable was as life-changing for rural communities as smartphones were decades later. Families went from receiving one or two fuzzy, weather-dependent TV channels, or none, to high-quality cable with hundreds of selections. Aaron had assembled an excellent organization with deep engineering and service experience that quickly won contracts for several Midwest towns.

However, they lost a few heartbreaking decisions to companies without experience in some towns because the companies or their owners had personal or business relationships with city decision-makers. Astonishingly, some city authorities were investors in Aaron's competitors but didn't disqualify themselves. Local citizens were as shocked as Aaron when city officials selected unproven companies without a track record over Aaron's company.

Aaron, though young at the time, was wise. He didn't want a reputation for suing prospective clients. More importantly, he knew that a quality of wisdom is to see life as it is, not as we would wish it to be, and adapt to reality. Early stage start-ups can't waste time chasing deals they will never get. Aaron began to prequalify prospects to avoid wasting time. His cable company went on to be wildly successful.

CHAPTER 3

Natural Law

"The ancients knew something which
we seem to have forgotten."

—ALBERT EINSTEIN

WHAT IF LIFE ONLY SEEMS RANDOM AND CHAOTIC but is instead a symphony with notes and movements you can understand and use to make wise decisions?

The natural laws that govern nature and the universe also organize economies, commerce, societies, relationships, and human psychology. Some natural laws are relevant to every challenging decision.

After reading this chapter, you will see world affairs, relationships, and yourself differently through the lens of natural law. Learning as many laws as possible sharpens your perspective of reality. By becoming familiar with natural law, you will significantly improve your decision-making.

NATURAL LAW DEFINED

The *Oxford English Dictionary* provides a near-perfect definition of natural law:

1. a body of unchanging moral principles regarded as a basis for all human conduct. "An adjudication based on natural law."

2. an observable law relating to natural phenomena.

"The natural laws of perspective."[1]

Humans are hardwired to innovate and advance human potential, civilization, and quality of life. Since ancient times, wise people have recognized that humanity and the natural world are governed by laws that result in better lives and societies when understood and respected. Mathematicians, scientists, inventors, physicians, and laypeople discovered natural laws and practical applications that led to the modern world we enjoy and too often take for granted. **Understanding and cooperating with natural law is a fascinating, quick, and effective way to make better decisions and live wisely.** Your explorations will provide expanding lifelong insight that few people today ever grasp.

The Fibonacci sequence and the golden ratio are thirteenth-century Italian mathematics discoveries demonstrating how applying natural law can explain and improve life. When the wealthy patrons of Pisa attempted to create a city that would be the wonder of the world, they were stymied by limitations to the height and scale of buildings they could construct. A local fellow nicknamed Fibonacci published a book based on ancient Indian and Arabic mathematics, providing solutions for building construction to previously impossible heights. As time passed, his work also explained seemingly unrelated natural phenomena, including the number of petals on a flower, the patterns of hurricanes, and the proportions of the human face. Even today, people who learn about the Fibonacci sequence and the golden ratio are astonished by its many applications in nature. Equally fascinating is the 2021 discovery of the "power cascade," also called the "power cone." It explains how sharp structures such as teeth, claws, horns, beaks, and more develop in animals with a precision that today's most advanced machines would struggle to duplicate.[2]

Natural law phenomena such as these are found in trees, plants, oceans, animals, and life's building blocks, such as genes, cellular structures, and subatomic particles. We can understand and apply such laws to improve our everyday lives.

NATURAL LAW IS AVAILABLE FOR EVERYONE TO DISCOVER AND USE

Albert Einstein, psychologist Carl Jung, astrophysicist and author Carl Sagan, and author Joseph Powell, among others, advanced a theory that natural law is found in music, art, inventions, and symbols like mandalas. Tribal leaders passed down the essence of natural laws in stories, parables, and colloquialisms centuries before the laws were scientifically discovered and proven. Jung and his students, for example, discovered archetypal human patterns and natural law in ancient fables, fairy tales, and proverbs. Examples include "A stitch in time saves nine," "What goes around comes around," and "What goes up must come down."

Perhaps someday, a natural law Rosetta Stone of sorts will be discovered that explains many of today's mysteries through a few hundred or thousand natural laws expressed mathematically and in other forms, like the Fibonacci sequence and power cascade. If that ever happens, human potential will accelerate even faster than it is today.

"Nature . . . never transgresses the laws imposed upon her."
—GALILEO[3]

You do not need to be a professional with advanced training to understand and apply natural laws. Science, philosophy, religion, and common sense sometimes align and diverge about the existence and interpretation of natural law but tend to coalesce around the principle attributed to an unknown source: **"Most answers we seek lie in nature if we only see them."** This chapter focuses on practical applications of natural law rather than scholarly debates. You will learn about a few essential natural laws and their applications. Ponder them carefully. Understand their principles. Do your own research. You only need curiosity, sharp observation, and willingness to apply them when puzzled or troubled by world affairs, your job, relationships, or difficult decisions. Done well and patiently, you will develop a natural law application process tailored to your learning style and needs.

THE LAW OF CAUSE AND EFFECT

"The first rung on wisdom's ladder
is respecting cause and effect."
—SHARAD DESAI

In physics, Newton's Third Law of Cause and Effect is that for every action, there is an equal and opposite reaction. In Eastern philosophy, this guiding principle is known as Karma. Colloquialisms and proverbs also capture its essence through sayings like "What comes around goes around," "You get what you give," and "Pay it forward." The law of cause and effect is one of the most straightforward natural laws to understand but often the most difficult to live by. Consistent application of cause and effect will largely define your quality of life.

We teach children cause and effect so they do not harm themselves or others:

"Don't run with that sharp stick."

"Don't point that sharp stick at anyone; it's dangerous!"

"See what happened when you ignored me! You are in timeout."

Children grow up respectful, responsible, and able to manage cause and effect if parents and other elders do their job well. Once adults, they respect the law of cause and effect instead of expecting society, its laws, the boss, or some other authority figure to force compliance. They live by a self-imposed standard captured in another proverb: "Character is what you do when no one is looking." We can assess our respect for cause and effect in the same way. Disregarding cause and effect always leads to suffering and heartache, whether in the macrocosm of nations and societies, the microcosm inside us, or the middle ground of relationships, careers, and business.

When I speak to groups, I sometimes ask audiences if they are familiar with the law of cause and effect. Everyone recognizes it and agrees that it is as consistent as gravity. But when asked about belief in Karma, typically half or less are believers. Audiences are invariably surprised by the heritage of cause and effect. Newton documented his third law in 1687. Karma was first reported in the first century BCE. Predating both by millennia is the Hindu teaching that the universe results from all previous actions.

When I ask the audience who among them complies with the law of cause and effect without exception, few hands go up. I ask why we never defy gravity but routinely defy cause and effect. Two answers are typical. First, defying gravity has immediate consequences, but the "effect" of denying cause and effect may be delayed or never witnessed. One person suggested that unpleasant happenings without discernible cause may be Karma's payback. Second, we may be unconvinced of the certainty of cause and effect, hoping that Karma is a lottery instead, especially when our reason to defy Karma is pressing. I then restate the law based on audience comments: for every action, there is an equal and opposite reaction, maybe . . . and only if witnessed. Audiences laugh at the capricious modification and get the point that cause and effect is a predictable law, not a lottery or one that requires witnesses.

Playing games with cause and effect is morally inconsistent with our self-image and self-imposed character expectations. If we believe that our lives and how we live them matter, we are morally obligated to accept responsibility for effects where our actions are the cause, irrespective of witnessing them. The excuses above mirror those of career criminals violating human-made law. How do we condemn their violations while excusing our own?

A lively discussion of cause and effect and Karma is a fun and informative parlor game for friends, family, and dinner parties. Don't be surprised by childlike naivete in adults, even though it is the natural law that most impacts us. **A wise person accepts that effect will always follow cause in some form, at some point, observed or not.**

Another misconception of cause and effect is that adversity is avoidable by living a kindly or saintly life. We likely benefit from the positive effects of a life where we did no harm because we originated positive causes, but adversity is inevitable.

You align with Newton's Third Law by doing your best to be your best and leaving the world and the people you meet a little better than you found them. Your every action affects your life and those you touch. That impact then ripples outward to people you will never meet through infinity. If you accept and live out that responsibility, your life will have greater meaning, and your potential will unfold sooner.

THE LAW OF POLARITY (OR THE TENSION OF OPPOSITES)

"But there is no energy unless there is a tension of the opposites; hence, it is necessary to discover the opposite of the conscious mind."

—CARL JUNG

The Law of Polarity, or the tension of opposites, is the most useful law for personal growth, second only to cause and effect. When I first learned of it in my early thirties, I responded, "Why didn't someone teach me about this in high school!" While it may be difficult to grasp initially, its impact is universal and significant.

Applying the Law of Polarity is one of the most effective methods for wise decision-making and developing an accurate perspective of the world and ourselves. It explains how our internal growth mechanism leads us to potential and wisdom. It provides a new perspective on tension, conflict, and adversity in your mind, relationships, business, and even between nations. The Law of Polarity explains the acrimonious divisions characterizing world society today and why such times come and go. With these insights, you can become a teacher and healer of divisions in the circles of your career, friends, and family. You can become a positive force for change, a pacesetter, instead of a helpless bystander.

OPPOSITES CREATE THE NEED FOR DECISIONS; TENSION CREATES WISE RESOLUTIONS

The tension between opposites generates creative energy for ideas and resolutions, including one that will emerge as best in time. Tension acts as a catalyst, challenging us to create something—a better world, better relationships, our best self. The law of polarity is a foundation of solar systems, relationships, democracy, and legal systems with a universally applicable decision framework.

A wise friend noted that the law of polarity is demonstrated in how physical strength is achieved. Weight-bearing and intense aerobic exercises create tiny rips or tears (microtears) of muscle fibers that initiate the body's repair process. The muscle grows back stronger than before. Repeatedly subjecting muscles to this natural rip and repair process causes them to become defined and strong, ultimately reaching their potential. This example of opposition creating an uncomfortable tension that results in growth provides the answer to another of life's unavoidable questions: *Why is life so hard and sometimes even seemingly cruel?* The answer may be what athletes often say: "No pain, no gain."

Opposites are evident in how we experience life: day and night, heat and cold, up and down, right and left. Our struggle to mediate and balance opposites in personality traits, beliefs, and actions defines our individuality. We attempt to balance the eternal and temporal, Soul and Ego, living for others or ourselves, accountability or compassion, etc. For example, the tension between living for ourselves or others is not one or the other or bouncing back and forth erratically, but in living for ourselves and others by mediating each factual situation to do justice to both.

We noted above the Ego's cognitive distortions of reality. One, called dichotomous thinking, is vital to understand when framing decisions. Dichotomous cognitive distortion is "either/or," "black/white" thinking. It causes us to judge polarities as superior or inferior to the other. In the law of polarity, opposites are neutral, functional, or constructive; neither is good nor bad or of greater or lesser value. Nature's opposites produce the energy for growth or forward movement. Unequal poles would dissipate the tension holding them in place and the catalytic energy required for creative resolution. You will be surprised how difficult it is to see opposing poles without dichotomous thinking. Try it as an experiment.

Some believe that treating opposites as balanced or neutral suggests that evil is on par with good or that there is no good or evil. As humans, we experience people and situations as bad or evil. That causes us to deny, destroy, or ignore the opposite polarity rather than recognize the tension as the means to a creative solution. Each pole in the tension, including evil, contains something for individuals or society to learn. In his book *Man's Search for Meaning*, author Victor E. Frankl explains how he discovered life's meaning and the indomitable spirit needed to defeat evil as a Nazi extermination camp prisoner. His story of wisdom's victory over unspeakable evil is a timeless gift to society.[4]

Most of us have or will face a situation that causes us to demonize another person or situation—an ugly divorce, a predatory boss, an unethical competitor, or a nasty neighbor. Sometimes, we later look back and discover the embedded lesson. If not, we will likely repeat the situation in a new factual form. The law of polarity teaches us to look for the lesson in opposition, even if it seems repulsive. **Society is divided** today because advocates foolishly reject the truth in opposition necessary for the tension of creative resolution. Problems recur in individual lives and societies if polarities remain unresolved by creative solutions.

This process that turns tension into creativity may seem theoretical, but observe it at work around you. A rubber band lying in your hand is inert. Stretched under tension, it binds things together. Stretched taut between two fingers, pulled toward you, and released, it creates directional energy, as any accomplished ten-year-old paper-wad warrior can prove. Stretched too far, the band breaks. This silly rubber band demonstration illustrates tensions inside you and in your career, relationships, society, and between nations.

The law of polarity is the foundation for representative democracy. It can be seen at work in the design of the judiciary. The tension between the opposing parties produces wise, creative decisions that honor precedent while adapting to new problems at a pace most of society can support. Your personal growth follows precisely the same principles. Today, the acceleration of change challenges you and society to adapt without breaking the rubber band.

INDIVIDUAL RESPONSIBILITY

"If you want the truth to stand before you, never
be for or against. The struggle between 'for' and
'against' is the mind's worst disease."
—JIANZHI SENGCAN, CIRCA 700 CE

Today's worldwide phenomenon of extreme polarization is the "struggle between 'for' and 'against'" that Jianzhi Sengcan spoke of thirteen centuries ago.[5] It illustrates a societal and institutional lack of wisdom of being unaware or failing to understand and apply the law of polarity. Blind advocates and rabid extremists reject natural and societal laws and norms of civility. They bully and demonize the opposition and rationalize barbaric acts in a charade of doing good. They accomplish nothing because their actions and reactions only strengthen their opposition; for every extremist action, the law of cause and effect creates an equal and opposite reaction. The fanaticism around winning, which derives from the Ego, dims the creative energy that could otherwise lead to a wise outcome. The potential for justice, moderation, and compromise remains outside of our reach. The cycle is self-perpetuating.

Escalating extremism strips people of their humanity, encouraging them to abandon healthy tension and compromise. That releases the destructive, vicious collective Ego and Shadow, the worst parts of ourselves. History proves that reckless extremism, when normalized en masse, produces no winners. Extremists and blind advocates inevitably fracture into warring factions that destroy each other, as in the French Revolution's guillotined extremist leaders and communism's endless purges. The silent, disgusted minority awakens to an unsafe society whose direction they no longer influence. Ultimately, undemocratic systems fail only after exacting horrendous tolls because they violate natural law, producing inadequate, inconsistent, untested decisions.

Extremism is partly fueled by the twenty-four-hour news cycle's alarmist emphasis. Degrees of extremism exist in every country, institution, society, and family. Moderate, responsible individuals are torn and unsure how to feel or what to do. Societies that allow this process to degenerate without checking it create the worst of times. When we read about how extremist societies begin or see them portrayed in a movie, we ask, "How could they allow that to happen?"

Healthy alternative views enlighten and elevate controversial topics and advance progress through tension-reducing resolutions over time. A world without polarities and tension is neither possible nor desirable because innovation and creative solutions would dry up.

You can take heart that dark times come and go for you personally and for humanity. The law of polarity repeats patterns in each of us and civilization; sometimes, those lessons are painful and tragic. Whether we like it or not, this is an age of increasing democratization and individual responsibility. We can't blame "those people" without looking in the mirror. We will be complicit in ushering in another age of fools unless we reject blind advocacy, extremism, and anyone who supports them, even if their cause, ideology, or leader is appealing. We must each be rational voices committed to peaceful civility and creative solutions. We must each choose to be wise or foolish. If we choose wrongly, future generations will ask, "How could they have let that happen?"

The law of polarity applies to our lives just as it does to world affairs. To experience more freedom, joy, and love, we first must accept hardship associated with the tension of the opposites. Rather than resisting tension, practice taking people and situations as they are, not as you wish them to be. Use tension as a force that helps you better understand yourself, other people, and circumstances in your life and the world. Understanding the law of polarity can help turn acrimonious division into healthy, productive tension and creative solutions in your orbit. You can be the pragmatic optimist with hope and optimism who explains polarization's purpose and beneficial outcomes. Embracing the law of polarity and tension of the opposites as a creative evolutionary engine inside us and the world can be your cause, a big step toward mastery of the art of living well.

Use this QR code to learn more about the law
of polarity and other natural laws.

THE LAW OF CONNECTEDNESS
(OR INTIMACY)

"When something vibrates, the electrons of the entire universe resonate with it. Everything is connected. The greatest tragedy of human existence is the illusion of separateness."
—ALBERT EINSTEIN

Intimacy connects us to everyone and everything. We can sense it even when we calmly look at the night sky. Intimacy is a little-understood transformative force that accelerates development to our potential. You can use it to connect to your Soul's timeless wisdom and inspiration, deepen relationships, love unconditionally, become free of possession and envy, and advance toward potential and wisdom by simply changing priorities from being liked or loved to embracing intimacy. **Beyond existing, there is love. Beyond love, there is intimacy. Beyond intimacy, there is everything.** Intimacy is a natural law inseparable from the art of living well. It is so important that an entire later chapter is devoted to its practices.

THE LAW OF RHYTHM
(OR OPTIMAL TIMING)

Time is a mental construct for earthly living. It exists only with change or an event. Einstein's Special Theory of Relativity states that the rate at which time passes depends on your frame of reference. Scientists and mathematicians disagree about how time may work in infinity or if it exists at all.

Why do we have ups and downs in emotions, moods, and outlook, as if governed by unseen tides? Why do life-changing events come out of nowhere for no discernible reason? How do we explain worldwide simultaneous inventions, designs, and inspirations, especially before the ships of the fifteenth century connected the continents? Why did artificial intelligence manifest as the most significant productivity tool in history just as population decline creates workforce shortages? Multiple proverbs capture time's curious nature about earthly events:

"Everything in its time."

"When the student is ready, the teacher appears."

"There is no such thing as an original idea."

"Man plans, and God laughs."

These sayings capture the essence of the law of rhythm, whereby events occur in their "right time." This idea may initially seem at odds with the law of cause and effect, but the two laws are symbiotic. Something occurs behind the curtain that science and technology cannot yet explain. It could be cause and effect functioning on infinity's timeline instead of Earth's. Perhaps the timing seems weird because time only exists in our minds.

The contemporary colloquialism "go with the flow" suggests that we should accept that much of life remains unpredictable, with a cause and effect we cannot delineate. Innovation, more than anything else, is changing our twenty-first-century lives. It is the product of inspiration, the arrival of which we can influence but not control. Unpredictability is predictable, so expect "black swan" events with severe consequences that seem likely or evident in hindsight, like the car that rear-ended you or the COVID-19 pandemic. But there will also be white swan events, such as an unexpected new friendship, rapidly developed COVID-19 therapeutics and vaccines, or an invention that improves your life.

Greet every new positive or adverse development as a puzzle piece, increasing comprehension of the inner and outer worlds comprising your life. Instead of responding to change with "Now what?" welcome it with "What can this teach me?" Record each unexplainable or surprising development, including inspirations, aha! moments, and ideas. Initially, they may seem random, irrelevant, minor, or nonsensical. Explore their meaning and how they might come together in a pattern or direction. You will discover that inspirations from months or years past can merge into an insight that explains how the world works or makes sense of your life in ways you never previously imagined.

THE LAW OF DIVERSITY (OR STRENGTH)

Nature's endless creativity is evident in its evolution and diversification. According to the Carnegie Science Earth and Planets Laboratory, "complex natural systems evolve to states of greater patterning, diversity, and complexity." Evolution is not limited to life on Earth; it also occurs in other massively complex systems, from planets and stars to atoms, minerals, and more.[6]

Life secures a foothold to grow, diversify, and evolve even in the most unlikely and forbidding places. Diversity strengthens life, while sameness weakens it, as overbreeding in animals has repeatedly proven. The Law of Rhythm (or Optimal Timing) and the Law of Diversity speak to an ongoing process of perpetual change in everything, everywhere. Nothing is static or the same from one moment to the next. By the time someone finishes saying, "People never change," all eight billion people on Earth are different.

VISUALIZING DIVERSITY

The previous chapter explained the bell curve, or probability distribution, to visualize individual differences across large populations or all of humanity.

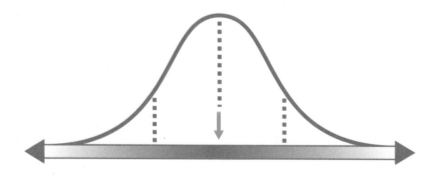

A bell curve is a common type of distribution for a variable, also known as a normal probability distribution, where the standard deviations from the mean create the bell shape.[7] Any symbol greatly oversimplifies diversity's complexity, and mathematicians sometimes debate the accuracy of the curve. Still, the bell curve functions as a useful reminder that diversity is continuously at work throughout the macrocosm and microcosm.

A flat, left-to-right bell curve fails to demonstrate what happens when extreme opposites share qualities, a phenomenon that has long been recognized.

Proverbs: "There is a thin line between love and hate."

Politics: Extreme left and right political spectrums both advocate minority dictatorships.

Psychology: Codependence can arise from unions of extreme opposite personalities.

Perhaps we think of the bell curve as a single-dimension, left-to-right image on a flat plane, when, in reality, it forms a mirror image or a sphere.

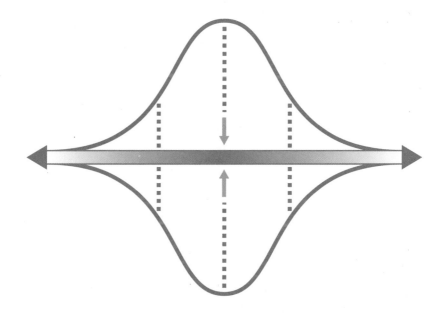

Diversity has an opposite: the Law of Connection or Intimacy. Things are diverse but connected by the common seed of creation.

REJECTION, CONFUSION, AND INCONSISTENT TREATMENT OF NATURAL LAW

Everyone accepts natural law to some degree to survive. We walk outdoors, trusting that gravity will prevent us from floating away. We rely on blue skies for a day at the beach and apply sunscreen to protect us from radiation. Life would be much easier if we consistently complied with all natural laws. Philosopher and theologian Saint Thomas Aquinas explained our choice: "Since human beings are by nature rational beings, it is morally appropriate that they should behave in a way that conforms to their rational nature."[8] Yet . . .

Aesop's fable of the ant and the grasshopper, written in the sixth century BCE, emphasizes that there is a time for work and a time for play. The ants store grain for the winter, while the grasshopper fiddles, choosing immediate gratification over planning for tomorrow. He dies, and the ants survive. This ancient, consistent, straightforward story is a lesson many of us have yet to learn.[9] Life expectancy has been expanding, and birth rates have been falling for over fifty years worldwide, creating an aging population that lives longer and has fewer young people to subsidize pensions and healthcare. Yet the average savings of people approaching retirement years in the United States is only $255,200.[10]

Children's survival depends on parents and other adults teaching cause and effect ("Look both ways before you cross the street! Don't touch the stove; it's hot! Never run with scissors!"). Adolescents and young adults rely on wiser people to learn respect for cause and effect. Unfortunately, some adults are poor role models. Understanding the reasons behind our disregard for natural law may help us to change and set an example of wisdom for young people. Following are a few reasons.

DEVALUING LIFE, STARTING WITH ONE'S OWN

We must consciously accept life as a gift or we will devalue it by elevating immediate gratification above future consequences, endangering ourselves and others.

EQUATING NATURAL LAW WITH PREDESTINATION

Some people believe life is predestined and that our behavior and beliefs do not affect how well we live. Life has patterns but there is little evidence that it is predestined. Even if it were, it would be impossible to guess how, when, and where. Predestiny avoids responsibility for our own actions, a form of fatalism that devalues life and our potential to live wisely.

THE ILLUSION OF CONTROL

Unlike those who believe life is predestined, others believe it is chaotic and up to humans (especially themselves) to bring order. Today, we have more predictability than ever in history because science has explained more natural laws. But believing that we can absolutely control our destiny is a cognitive disorder or illusion created by the Ego. Life is neither predestined nor within our absolute control, but we can increase predictability by respecting the natural laws we know and understand.

CONVENIENT INCONSISTENCY

Some people selectively reject or ignore natural laws when emotions, instincts, biases, lack of perspective, or their Ego overwhelm their capacity for wisdom. Few people deny gravity to jump off a ten-story building but deny cause and effect because the consequences are delayed or unseen.

APPLYING NATURAL LAW IN DAILY LIFE

DENIAL AND THE LAW OF CAUSE AND EFFECT

Denial or suppression of memories, emotions, events, and thoughts is ineffective because of the Law of Cause and Effect. The more you try to deny a thought, feeling, or attitude, the stronger it becomes. The longer you deny natural law, the more it impacts you and your life. Yet denial is the first reaction of most people. Many live their entire lives futilely denying and suppressing what they are afraid to confront.

THE LAW OF POLARITY INSIDE US

Publishing the Potentialist book series for an audience beyond family and friends was never my plan. When early draft readers convinced me to consider it, I remained torn. After years as a high-profile CEO, I enjoyed my anonymity and had little appetite for being a public figure again. I also questioned my qualifications to communicate such a complex subject to a large audience beyond family and friends.

I used the law of polarity to discover a wise resolution. Polarities, like icebergs, often have more below the surface than above. The surface decision was, "Do I publish to the general public or limit publication to family and friends?" My difficulty answering that question suggested I needed to dig deeper. The polarities of tough decisions are almost always temporal/eternal, collective/individual at their core. Humility was the underwater iceberg of my desire for anonymity and my author qualification concerns. Its opposite was individuality. Armed with neutral opposites, a creative resolution came within hours: "Always be humble but honor your gifts." The creative resolution to the opposites honored the humility I held about my author skills and my desire for anonymity. The resolution equally honored my half-century research and practices in human potential worthy of sharing if I could do so clearly in everyday language. The resolution has guided my Potentialist work as an author and speaker since.

CAREERS AND THE LAWS OF POLARITY, DIVERSITY, AND INTIMACY

The most effective leaders and professionals I have known over a long life demonstrated an understanding of natural law, including highly skillful CEOs, executives, salespeople, attorneys, and negotiators. They saw opposing parties and issues through a lens of fairness, balance, and empathy. They understood individual differences while recognizing predictable archetypal behavior and situations. They recognized trust as the door to intimacy and treated their opposition with civility, respect, clarity, and transparent intentions.

CREATIVE DURABLE SOLUTIONS

Dr. Lisa Fischer advises people to visualize tension or conflict between them as an opportunity to collaborate on mutually satisfactory solutions rather than something to be avoided. She suggests they visualize the tension as a physical object with polarities and a spectrum of possible solutions. She asks opposing parties to sit side by side instead of opposite each other. The issue is then constructed on a table in front of them, on paper as bullet points, drawings, or imagined structures. She encourages the parties to seek increased intimacy and mutual empathy through their work together, not simply conflict resolution.

Dr. Fischer's method is genius. Moving the issue, its polarity, and tension metaphorically onto a table moves the parties from being positioned as opponents to resolution partners with the potentially high reward of increased intimacy. Describing the issue together in detail typically finds common ground. Creativity rises to more significant challenges, resulting in a mutually satisfying resolution. Dr. Fischer's approach applies to any personal or career tension or conflict.

DIVERSITY AS A BUSINESS STRATEGY

Seeing through the bias that lies in assumptions is an area ripe for innovation. Virgin Enterprises founder Richard Branson created a worldwide business empire by providing exceptional products and services to markets and customers neglected by industry leaders.

Silver Sneakers became one of the most successful fitness plans in the United States (the average participant age is seventy-eight) because entrepreneur Mary Swanson defied the late 1990s conventional wisdom that "older adults won't exercise." The bell curve argued that some did or would, especially if a program was explicitly designed for them. Mary partnered with me and my son Hugh to expand nationally. Together, we achieved substantial growth and financial success as millions of loyal seniors became Silver Sneakers members. We also had a great experience building something of value and meaning.

SEVEN PRINCIPLES FOR APPLYING NATURAL LAW

1. Observe natural laws at work inside you and in the world around you. Use natural law to define and solve problems and make

difficult decisions. It will make your decisions easier, faster, and wiser and will soon become a habit.

2. There is always an effect to cause, even if we never see it. Accepting responsibility for an effect we may never see is a character-defining moral choice. We remain unwise and unable to accept ourselves if we deny responsibility for a cause to an effect.

3. Understanding archetypal situations and behavior patterns and knowing when they apply to decisions can predict likely outcomes, develop broader perspectives, and shape wiser decisions.

4. Diversity strengthens species, ecosystems, and nature. Differences may seem unfair or even cruel from a human perspective, but they serve nature's function. Diversity is mediated by intimacy or connectedness with all things.

5. Apply the law of polarity to all complex decisions. Every tough decision has definable opposites and a spectrum of creative resolutions when polarities are clearly defined, and neutrally valued.

6. Creative resolution doesn't always work in your desired or required timeframe. Allow the Law of Rhythm to work for you. When possible, wait for an idea, innovation, or a change in external circumstances to arise instead of forcing a solution. Your patience will be tested, but it pays off.

7. Use the tension of opposites to increase intimacy instead of diluting or destroying it. Strangers can become friends. Enemies can become allies. Friends can become family. Each of us can become more whole.

BROADER PERSPECTIVES, USEFUL TOOLS

Understanding and applying natural law improves decision-making and perspective of the world around us. It is a significant career and life advantage that will become essential in the decades ahead.

WISDOM STORY:
FROM CONFLICT TO CREATIVE SOLUTION

Eugene Hinkle was a fascinating man. He grew up near Baltimore. After graduating college with an engineering degree, he joined the paratroopers in World War II. He began the war as a second lieutenant and ended it with a field commission to a higher rank. Postwar, he became involved in the rapidly developing computer technology field. He rose to senior management with the aerospace company Martin Marietta before joining Blue Cross and Blue Shield of Indiana as its chief information officer. He was my boss from 1976 to 1980 and mentored me to succeed him.

Eugene, or "Gene" as I knew him, taught me, among other things, how to look and act like an executive undaunted by the other people in the room. My favorite wisdom story about Gene was his skill at diffusing and redirecting tension toward a creative solution. Among his many talents, Gene was a US Civil War buff. He could describe most major battles in detail and had visited many battlegrounds. He said he used Civil War tactics leading his paratroopers in World War II.

One day, we had an important meeting between our information technology leadership team and a group of internal clients about a new system we were developing. The project wasn't going well. Nerves were on edge; tempers were short. Soon, a full-fledged argument was underway. As the meeting chair, Gene attempted to calm things and get back on track, but to no avail.

With everyone arguing, Gene quietly went to the blackboard and began to draw what appeared at first to be a football play diagram with X's and O's. But then he drew buildings labeled "Courthouse," "Joneses' farm," and such, and squiggly lines labeled with the names of creeks and rivers. Slowly but surely, the argument ceased as the executives in the room became fascinated by what the heck Gene was doing. When he sensed he had everyone's attention, Gene launched into a fascinating lecture on a specific Civil War battle. We were spellbound.

In the end, he had us in the palm of his hand and said, "Do you think those officers about to lead their men in battle would have behaved like we were a few minutes ago?" No one said a word. Gene said, "I didn't think so. Now, let's get back to work and finish this project." He masterfully framed issues to respect each point of view. Resolutions came surprisingly easy. The project went on to be a success thanks to his wisdom and intuitive use of the Law of Polarity.

CHAPTER 4

The Inner World of Your Psychological Body

"You will never have a greater or lesser dominion than that over yourself."
—LEONARDO DA VINCI

WHAT IF EVERYTHING YOU NEED to achieve your potential, become wise, and make better decisions to live well is already inside you, waiting to be released?

YOUR PSYCHOLOGICAL BODY OPTIMIZED

You can optimize your psychological functions, collectively called the psychological body, as easily and naturally as your physical body. You already perform many of the necessary practices unconsciously and only need to make them conscious and deliberate. This chapter doesn't try to make you an amateur psychologist. It provides enough information to optimize the psychological body instead of it being a mystery.

Athletes strive for "peak performance," a composite effect of conditions and habits—physical activity, nutrition, sleep, quiet time, etc. This principle applies equally to the psychological body; you optimize performance by integrating its functions using simple, easy-to-learn practices. Your potential, the INME, becomes primary in consciousness, available to make wiser decisions.

THE EVOLUTION OF THE PSYCHOLOGICAL BODY

No one knows for sure how human psychology developed, but speculation can be useful as a framework for our psychological body and how it could be more effective with a little effort.

The Shadow may have emerged first as part instinct and part conscious, followed by the Conscience as an opposing function. The Ego may have developed next or in parallel as an outgrowth of the Shadow. The Ego was likely the first fully self-aware, conscious psychological function and may have been the only one for tens of thousands of years.

The first record of the Soul function appeared around 8,000 BC. It described the Soul as "God inside me (the Soul), but distinct from me (the Ego)." The true self (INME), as distinct from the Ego and Soul, also has ancient origins.[1] In the late nineteenth and early twentieth centuries, philosophers and early psychologists began researching and writing about the self with varying perspectives on its origins, development, functions, importance, and future. Much of that research continues today.

The pursuit of our best self is on the rise, reflected through increased interest in books like The Potentialist series, social media traffic, popular public speakers, and programs like mindfulness. Humanity can now create near-human-like automation. We must also have the wisdom to establish the proper roles and boundaries for its use. Each of us must understand and preserve what makes humanity distinct, which requires knowing our true self INME and an intimate connection with the Soul. The decline in religious worship makes individual Soul connections essential.

Adaptive change to our psychological body for a New Reality seems destined, likely as a step change rather than gradual. The INME could replace the Ego's domination of consciousness during the twenty-first century. You are living at the beginning of this historic change and can be an early adopter, benefiting you, your family, and humanity. Developing your INME and the wisdom it offers should be your highest priority to adapt to the New Reality.

THE PHYSICAL AND PSYCHOLOGICAL BODY KNOWLEDGE GAP

Your psychological body impacts happiness, relationships, and success more than your physical body, but most people know little about it or how to use it to live better. By our early teens, we know our physical body well from observing, experimenting, and learning from teachers, parents, peers, books, movies, and the internet. We spend considerable time daily grooming, exercising, feeding, and caring for our physical body.

But few adults know much of anything useful about the psychological body and do little or nothing to nurture its health and growth. Even so, it reminds us that it's there: a racing heart when we're nervous, an upset stomach from worrying, unrestful sleep when we feel stressed, and increased energy when we get excited. We live inside our psychological body from birth as it grows, evolves, and dies like our physical body. But the psychological body, unlike the physical, is self-aware and exists unto itself. The physical body is an empty shell without the psychological. The knowledge gap between the two bodies is striking, archaic, self-limiting, and unnecessary.

ANCIENT ATTITUDES IN A NEW REALITY

Before modern psychology's emergence, most people in Western cultures treated the operations of minds, hearts, and souls as a mystery. Impaired cognitive function and odd and nonconformist behaviors were often treated as curses, and later as diseases of the mind.

Even today, some cultures believe avoiding psychological issues to be the best course. That makes them more vulnerable to negative effects of the Ego and Shadow. Even in developed countries, some individuals refuse to seek medical exams or a diagnosis, fearing what they will discover. Many more avoid psychological counseling or therapy for the same reasons.

Fortunately, modern psychology has expanded far beyond only treating mental illness to tapping individual and collective human potential, at the same time that medicine is following a similar path from focusing on curing disease to optimizing health.

RECOGNIZING YOUR TRUE SELF INME

A first step to understanding and using your psychological body is to recognize thoughts and voices distinct from your own that are psychological functions, especially the Ego. The Ego dominates your consciousness today, so much so that you can't quite see yourself as separate from it. But the Ego is a psychological function constructive today only in its proper role. It is not you and should not define you.

The Ego's primary mission is survival. That mission was appropriate for most of human existence when physical survival was an everyday fight. But change accelerated faster than our psychological body adapted. Now a more civilized, sophisticated, wise you is required to address the real threats and opportunities of the New Reality. The Ego is like the relative who has overstayed their welcome and needs to get their own place.

It resists psychological functions that compete for your consciousness. You might experience this when you meditate, engage intimately with a loved one, read this chapter, or try to sleep. All these pursuits allow you to seek out the Soul. Instead of getting out of the way so you can access the Soul more easily, the Ego puts unwanted thoughts in your head, denies the Soul's existence, discourages self-discovery as a waste of time, and causes you to become impatient with loved ones who want your full attention. The Ego always makes you feel that something else is a higher priority, such as work to do. Everyone struggles with the Ego's endless demands. It will get worse as the pace of change accelerates.

You likely experience situations frequently when your Ego won't be quiet so you can concentrate on something more important. These annoying situations are opportunities to differentiate the real you/INME from the Ego. If the Ego puts unwanted thoughts in your head, who is trying to concentrate, and who is interfering? If one voice won't let you go to sleep, who is trying to sleep? Who is talking, and who is trying not to listen? The Ego is the troublemaker in all these examples, and the one struggling to do what you really want is the real you/INME. Until now, it has likely been largely or totally unknown to you. Meet yourself and say hello.

Why should you have to fight to concentrate? I mean, who's in control here? Can you visualize and relate to how strange this is? Imagine if your left arm and hand suddenly began to slap your face autonomously! You would race to a hospital emergency room because your body function was out of control. But daily, the Ego psychological body function acts autonomously against your wishes. **Placing the real you/INME in charge is the struggle for your potential, becoming wise and fully human.**

LIVING IN AND RECONCILING TWO WORLDS

You may have heard or read that humans are both temporal (earthly) and eternal (infinite) creatures. Visualize two exactly opposite worlds inside of you, each with distinct characteristics.

Temporal (Earthly) World	Eternal (Infinite) World
Ego/Shadow	Soul/Conscience
Individual Interests/Aspirations	Other's and the Whole's Interests/Aspirations
Hierarchical	All Creations are Equal
Survival of the Fittest	All is Eternal in Some Form
Time is Everything	Timeless or Unfathomable Time
Diverse, Differentiated, Unique	Endlessly Creative
Appearance of Unchanging	Everything is in Motion and Changing
Familiar, Knowable, Comfortable	Unfamiliar, Unknowable, Uncomfortable
Power, Domination, Control	Unfathomable, Uncontrollable
Five Senses	Senses Beyond Five
Definable Beginning and End	No Beginning or End

The psychological body is a system of opposing forces, a reflection of the Law of Polarity described in Chapter 3. Functions like the Ego and Shadow evolved so humans could survive and prosper physically and materially on Earth. The Soul and Conscience connect us to timeless truths and laws so that every new generation and individual can carry forth an understanding of how things work beyond the physical plane and our five senses. Your psychological body is aware of its existence and acts accordingly. Like animals and automation, your physical body is not self-aware, so the Soul and Conscience prevent self-destruction or destruction of the species. Together held in balance, these psychological functions support our ability to succeed as temporal and spiritual beings.

Healthy tension between the temporal and eternal/spiritual worlds is the foundation of your psychology and the root of every tough decision requiring wisdom. Everything in the temporal has a compensating mirror image in the eternal: Shadow-Conscience, Ego-Soul. The creative product of the polarities is the INME. It crafts compromising resolutions between the conflicting functions. The INME with each use becomes wiser, more present, and able to keep the Ego and Shadow in their proper roles. The wiser you become, the more the law of polarity explains life.

The origin of the Soul/Conscience is debated today as it has been for millennia. Some insist it is the manifestation of God in individuals; others reject the Soul/Conscience's existence because they are hostile to spirituality. We do not have to agree on the origins of our psychological body to work with its functions. If you find the names of the functions troubling, substitute Individual Aspiration for Ego and Aspiration of the Whole for the Soul.

CHECKPOINT

Let's pause briefly and review salient points from previous chapters and this chapter thus far:

1. Your life and how you live it matters. Other unavoidable or existential questions must be answered if you are to direct the course of your life instead of stumbling through it.
2. You are subject to natural law constantly at work inside your head and in the outer world.
3. Understanding and applying the laws of cause and effect and the law of polarity are essential to wise decision-making.
4. To become wise and master the art of living well, you must assume responsibility for your psychological body and its highest achievement, the INME, the emerging real you, your *potential*.

You are not the functions of your Ego, Shadow, Soul, or Conscience. You are the INME.

5. Plant one foot firmly in both the temporal and eternal worlds and reconcile differences through wise decisions. From the temporal world, we find meaning in the dignity of self-achievement. From the eternal world, we find enlightenment from self-discovery and eternal truth. Both are essential to living wisely and well.

PSYCHOLOGICAL FUNCTIONS OVERVIEW

You likely have some, or perhaps a great deal, of familiarity with the Ego, Shadow, Soul, and Conscience but little or none with the INME. This chapter intends to expand your knowledge and make the functions more practical and usable.

THE SHADOW

"Out of your vulnerabilities will come your strength."
—CARL JUNG

DEFINITION

Psychological schools have varying definitions of the Shadow. Generally, the Shadow is a part of us hidden or repressed. For a layperson's use, this discussion is limited to how we experience the Shadow and can work with it to grow instead of resisting it.

ORIGINS

The Shadow is a part conscious and part instinctive function manifesting in different but recognizable ways. We consciously hear its distinct, unpleasant voice, and it sends physical messages when necessary. For example, imagine a 15,000 BCE hunter pursuing an antelope. His Ego consciously knows where they graze, and he intends to spear one to feed himself and his family. As he approaches the herd from behind a rock outcropping, the hair stands up on the back of his neck, and he instinctively freezes and watches a lion walk where he would have been standing. The Shadow's warning saved his life. That slightly evolved Shadow function alerts you today to possible threats on a dark street or a hostile reception at a business meeting.

FUNCTION

The Shadow alerts us to unconscious or unresolved, real, or imagined threats that make us feel vulnerable. It is the vulnerability we feel that activates the Shadow. Its job is to alert our consciousness and pester us until we accept and resolve the vulnerability; it plays hardball if we deny it. For example, the Shadow shows up when our defenses are down, such as before we fall asleep, during sleep, or if we awaken in the middle of the night. It uses harsh language, fear, condemnation, shaming, or frightening dreams to get our attention. A Shadow visit is never sought or pleasant. The Shadow relentlessly returns with increasing intensity until we no longer feel vulnerable.

The instinctive Shadow may also appear if a severe vulnerability is triggered. It can cause extreme instinctual, unconscious behaviors such as snapping, yelling, threatening, or even physically harming someone. This type of Shadow behavior can trigger Shadow responses in others. It occurs millions of times daily with tragic results. Carl Jung said of this type of Shadow response: "The best political, social, and spiritual work we can do is to withdraw the projection of our shadow onto others."

For all its unpleasantness and lethality, the Shadow, when utilized effectively, is a constructive teacher that reveals vulnerabilities that would otherwise unconsciously make us miserable, fearful, and anxious. The Shadow rewards us for making vulnerability conscious and resolving it by going silent. Then, it moves on to the next unconscious vulnerability that it detects.

WHEN AND HOW EXPERIENCED?

You can recognize the Shadow by its negative harshness and how it can make you doubt yourself or self-loathe. It may say, "Who are you to do X or Y? You will fail," or "No one likes you." The Shadow does not know if the vulnerability or its harsh language is true. It is a function whose job is to reveal your unconscious to you and is oblivious to truth and reality. It is bringing to consciousness your worst unconscious opinions and doubts about yourself. Unlike the Ego, the Shadow doesn't tell you how to protect yourself. Nor can it determine whether you are genuinely vulnerable; it can only let you know you feel exposed.

For example, you take a new job. Consciously, you are happy about it and eager to make the change. That night, you become restless before sleeping or awaken to worries and doubts voiced by the Shadow. It may taunt you with accusations of being a fool for leaving a safe, cushy job. This is the Shadow reporting the doubts that you consciously denied.

WORKING WITH INSTEAD OF AGAINST THE SHADOW

The Ego urges domination of our environment and others to be safe. The Shadow reminds us of where we feel vulnerable even if the Ego has convinced us all is well. An activated, intensified Shadow can cause you to react unpredictably as if you are a different person. Immediately upon acknowledging the Shadow's message, it will lessen its intensity.

The Shadow optimally utilized is a reservoir for growth. You can discover your unconscious or unresolved vulnerabilities and put them behind you. Don't react with dread or despair when the Shadow speaks, and don't take what it says literally. Instead, understand why and to what it perceives you to be vulnerable. Be grateful for its function to protect you. Acknowledge and accept responsibility for resolving the vulnerability.

CONSCIENCE

"Listen to your conscience. That's why you have one."

—FRANK SONNENBERG

DEFINITION

The *Oxford English Dictionary* defines conscience as an inner feeling or voice viewed as acting as a guide to the rightness or wrongness of one's behavior.[2] Almost everyone knows this function well and easily differentiates it from our true self. Activation of the Conscience becomes rare as the INME grows strong.

Unlike the Soul, your Conscience can trigger shame or self-judgment that can cause you to deny its message or ignore it. Similar to the Shadow, the Conscience is not shaming or judging you. It is bringing to consciousness the self-judgment you feel but are denying or ignoring.

Working with your Conscience is similar to the Shadow. Don't shy away from the message. Accept responsibility to deal with it consciously. Then, the Conscience has served its function and will move on to the next thing you are denying or avoiding.

FUNCTION AND EXPERIENCE

The Conscience is a messenger for the interests of the whole, beginning with family and extending outward. For example, imagine the hunter-gatherer discussed earlier becoming exhausted and famished. His hunger calls him to devour the entire small antelope that he killed. His Conscience reminds him of his starving wife and children. The hunter must choose between his personal interest and the collective interest of his family. As discussed, every wise decision is a creative resolution of tensions between the Ego/Temporal and Soul/Eternal. Eating enough to avoid starvation but saving most of the kill for his family wisely resolves the tension.

WORKING WITH INSTEAD OF AGAINST YOUR CONSCIENCE

Humans would be little more than beasts without a Conscience. Families might not have survived, and tribes, communities, and nations might not have developed. Laws would have been unenforceable without the Conscience's reminders of their importance to the whole. We should be thankful that we have a Conscience and listen closely and gratefully to its message. We should never be angry with our Conscience for the opportunity to decide wisely. The Conscience reminds us of past decisions we regretted, helping us avoid the repetition of unwise choices.

EGO

EGO DEFINED

The Ego in psychoanalytic theory is that portion (function) of human psychology experienced as the "'self' or 'I' in contact with the external world through perception, the part that remembers, evaluates, plans, and is responsive to and acts in the . . . physical and social world."[3]

The Ego exists inside you and in the collective Ego of groups that exhibit the same characteristics. This individual and collective quality is true of all psychological functions. You could spend a lifetime studying the Ego. Fortunately, that isn't necessary to change the Ego from decision-maker to adviser and from running your life as "you" to a psychological function in the domain of the real you/INME.

THE MISUNDERSTOOD EGO

The Ego is often demonized in contemporary discussions and media as a synonym for conceit and self-aggrandizement. The Ego is far more complex, a psychological function essential to our survival and ability to prosper in the temporal world. Grasping survival as its primary mission is critical to using the Ego effectively.

Confusion about the Ego arises because its survival drive manifests in modern life as aspirations for wealth, power, fame, appearance, social status, and the desire to achieve, build, and explore. The Ego can be deceptively lethal, using charm, illusions, and false logic in its single-minded pursuit of safety, including its drive to dominate and control your conscious mind. Without the Soul's counterbalance, the Ego has no ethic, no genuine interest in the best interests of others, and is incapable of giving or receiving intimacy or unconditional love. You can't love an Ego, and it can't love you. The real you can be wise, but not the Ego. It causes you to fear what is no longer a threat while ignoring new, more urgent threats.

It can ruin your life and rob it of meaning, but it is a powerful source of energy and direction in its proper role.

MEETING YOUR EGO

I was first informed that the voices and thoughts in my head were often not my own in a college psychology class. I was initially skeptical and suspicious that the professor was using an illegal substance. But then she asked questions that entranced the class: "Do you recognize a voice in your head that never shuts up, even when you want it to? Do you hear thoughts or say things unrepresentative of what you believe? Where do they come from?" After posing a series of questions like these, the professor turned the skepticism of our class into curiosity, experimentation, and acceptance of the Ego as a distinct function.

IDENTIFYING THE EGO BY ITS CONSISTENCY

The Ego's complexity is not easy to digest but is made easier by its consistency. All Egos behave much the same: yours, mine, and the collective Ego. All can quickly be identified. We've all heard, "He's in his head." That's Ego identification. The Ego's stereotypical thoughts and behavioral patterns help you know when you or others are in the Ego or the INME.

Below are quick-reference examples of stereotypical Ego motivations, thought patterns, and behaviors. With practice, you can add your own observations. Think of actions, attitudes, thoughts, or mindsets that you do not like about "yourself." Many or most are the Ego, not the real you/INME. You may see them or their origins in the lists below. Identifying the Ego's petty, controlling, aggressive, or embarrassing thoughts and behaviors differentiates it from the real you/INME.

Ego Survival Characteristics	Ego Aspirational Characteristics
Threatened by anything beyond the temporal world and five senses	Active manager of the temporal world
Builds defenses and demands action to resolve real, imaginary, or exaggerated threats	Aspirations can moderate the survival responses
Becomes increasingly aggressive based on stress and perceived threat severity	The busiest part of the psyche that can become hyperactive under stress
Demands solutions by nagging relentlessly, "You must do this" or "You need more"	Positively expressed as a need to plan, execute, decide, explore, know, conquer, achieve
Demands attention over rest, relationships, quiet, and Soul connection	Source of ambition, a positive effect of the need to control and dominate
Seeks control of entire life and environment and resists constraints, including natural law	Source of temporal meaning in the dignity of self-achievement
Can overreact dangerously and self-destructively	Seeks to grow and improve to differentiate and increase security as a species leader
Undesirable, embarrassing, or dangerous behavior such as jealousy, envy, lying	Seeks opportunity to enhance individual, tribe, and species survival and prosperity
Linear thinking and tunnel vision	Can react logically and rationally, but not wisely, which requires the INME
Unethical, selfish, and predatory behavior unless moderated by the INME	Ego thoughts can be quite evolved but revert to primal responses under stress

THE EXTREME EGO EXPERIENCE

As previously noted, unrestrained egos create self-destructive cognitive distortions. Because the Ego has been dominant in consciousness, many cognitive distortions have become societal norms. Consider the mindshare and life activity in service to these fictions and how they inhibit humanity from achieving its potential. Your life will improve immediately as you comprehend and reject these distortions.

1. *Control* leads to fear and paranoia since nothing is absolutely controllable in life.
2. *Possession* of people and things is futile because they were never yours in the first place.
3. *Power* leads to the illusion of being above laws, even cause and effect.
4. *Domination* of consciousness, people, and environment leads to resentment and rejection.
5. *Perception of endless scarcity* leads to insatiable demands for more.
6. *Perfection* leads to indecision, delayed or avoided commitment, and self-loathing.
7. *Suspicion* beyond prudence leads to an inability to trust or be intimate.
8. *Demand for certainty* results in fear when it cannot be achieved, which it rarely is.
9. *Impatience* results in immediate gratification instead of what is best and achievable.
10. *Overgeneralizations* base invalid conclusions on statistically insignificant samples.
11. *Magical or illogical conclusions* attribute effects to no proven or possible cause.
12. *Mind-reading* is a false belief in the ability to interpret others' thoughts and beliefs.
13. *Fortune-telling* is making predictions without evidence or logic.
14. *Emotional reasoning* equates emotions with reality.

15. *Dystopian or negative thinking* emphasizes the negative by ignoring the positive.
16. *Life as we want it, not as it is* ignores reality for what the Ego says it should be.
17. *Dichotomous thinking* is all-or-nothing reasoning that ignores the law oaf polarity, natural variability, and diversity.

EGOS EQUATE SAFETY WITH SUCCESS
INSTEAD OF MEANING

Ego Success	Soul/INME Success
Money	Meaning, Quality of Life
Power	Wisdom
Possessions	Intimacy/Relationships
Appearance	Character
Social or Professional Status	Leave the World and People I Meet Better
Always More	Living Well with Enough
Fame	Not about Me
Safety	Face Whatever Life Serves Up

PRIORITIZING DISTORTIONS AND TEMPORAL SUCCESS OVER MEANING HAS SEVERE CONSEQUENCES

1. *Lost relationships*: too busy working, controlling, possessing, envying.
2. *Excessive risks*: foolish bravado, reckless behavior.
3. *Identification with public persona* leads to feeling fraudulent with fear of discovery.
4. *Meaning that is defined by others* leads to feeling insignificant when it is lost.
5. *Opportunities missed* due to pessimism, indecision, and inability to trust.
6. *Obsessive and paranoid behavior*: attempting to control the uncontrollable.
7. *Pollution or denial* of memories, dreams, inspiration, and life lessons.
8. *Possession, jealousy, or envy* leading to victimhood, revenge, or criminality.
9. *Lack of self-discovery and awareness*: overestimating or underestimating oneself.
10. *Defeatism* to avoid effort and risk leading to stunted growth.
11. *Winning at all costs*, including using charm and false logic without concern for consequences.

PRINCIPLES FOR SUBSTITUTING YOUR INME FOR THE EGO

The Ego moderates when it is focused on actions or goals to achieve and when it is subordinated to the INME. In these circumstances, the Ego manifests as the "can do" we like and admire in ourselves.

1. As described above, the Ego's voice, thoughts, and actions are distinctive and easily recognizable in yourself, others, and the collective Ego.

2. Sometimes, the Ego feels like you because it has dominated your consciousness. But often, it feels like someone else or a part of you that you dislike.

3. The Ego is a limited, predictable psychological function. The real you/INME is much more. You can recognize that your INME is active in consciousness when you feel good about yourself, act honorably, make a wise decision, or balance your interests with the interests of others.

4. Speak your thoughts to the Ego as a function to incrementally strengthen the INME and consciously, deliberately differentiate it from the Ego.

5. Avoid judging, denying, or fighting the Ego, which increases its power.

6. The Ego's mission, value, and voice should be heard and honored, but it should not own your consciousness or run your life.

7. Survival always trumps the Ego's aspirational mission.

8. Connect with the Soul as often as possible to counterbalance the Ego. Use the practices in this book to aid connection.

9. Be patient. It may feel strange to differentiate yourself from the Ego at first. Stretch yourself and it will feel normal with practice.

10. Identify and stay centered in your INME once you sense it. You have achieved a life-changing developmental milestone when you routinely differentiate the real you from the Ego.

EXAMPLES OF SHIFTING CONSCIOUSNESS FROM THE EGO TO THE INME

Surprisingly simple shifts in attitude, mindset, feelings, and actions elevate your INME and put the Ego in its place.

You volunteer to coach your child's soccer team. Parents compliment you profusely for doing a good job. You feel your Ego swell with pride. Deflect it to your higher purpose. Say to the Ego in your mind and ideally to those complimenting you, "Thank you, but this isn't about me. These are courageous, coachable kids." Doing so engages your INME and disengages the Ego. Reflect for a moment after on how lousy and small you would have felt serving the Ego and how grounded you feel serving yourself and others equally, a defining quality of your INME.

A loved one attempts to converse with you after a hard workday, but your Ego retains control. They challenge, "Are you listening to me?" Don't be defensive. Admit that your Ego has you and won't let go, but they are your priority. Ask for a few minutes to collect yourself. Perform the calming exercises in Chapter 7, "Learning States," or go for a walk. Tell your Ego that you respect its need for attention, but you are needed elsewhere, so please go to sleep.

Invent techniques to subordinate Ego behaviors to the INME, as in these two examples.

SOUL

"As far as we can discern, the sole purpose of human existence is to kindle a light of meaning in the darkness of mere being . . . Who looks outside, dreams. Who looks inside awakens."

—CARL JUNG

SOUL DEFINED

The Soul is our personalized experience of timeless wisdom, natural law, and the aspirations for the whole (creation). It is a semiautonomous function that we can invite but not command to respond. **The Soul comes to us when it senses a genuine desire and intention for** an intimate connection. This may sound strange or uncomfortable; creation and infinity, which are qualities of the Soul, are unfathomably different from Earthly life.

For many people, denying or questioning the Soul's existence is easier than admitting fear of something they cannot understand or control. Others pretend to have a Soul connection when they meditate but cannot quiet their Ego to achieve it. They seek to deceive themselves and others because they are embarrassed that their Ego dominates them. Many people aren't fearful of the Soul but struggle to purposely connect the first time.

Helen Keller's words are instructive: "It gives me a deep comforting sense that things seen are temporal and things unseen are eternal." Thankfully, many people have a deep connection to the Soul, even if they never talk about it or don't fully understand it. They cannot imagine life without the Soul, and if you can get them to speak about it, they use terms like "bliss," "heaven," "peace," "ecstasy," "intimacy," or "unconditional love."

Having a Soul and connecting to it is one of the few defining characteristics of being human. It is the source of our aspirational drive to reach our potential. We feel fulfilled and successful only by cooperating with this drive. When intimate with the Soul, we become connected to its dimension: all that was, is, and ever will be. Yet the Soul's relationship is personal, loving, guiding, and perpetual. Its capacity to be both infinite and personal is its majesty.

ACCEPTING THE SOUL AS A PSYCHOLOGICAL FUNCTION

If you relate to the Soul easily and regularly, you will enjoy and breeze through the following few pages. If you reject or doubt the presence of the Soul, identify with it solely in its religious context, or cannot connect with it, please remain open to expanding your perspective.

Your challenge to connect with the Soul is understandable
The Ego's endless chatter must be quieted to hear the Soul. The Ego aggressively resists because it loses domination of consciousness once you routinely connect to the Soul and the INME. The Ego and collective Ego argue that you have no spiritual function or need for one. Scornful techniques are used to intimidate, including rationalizing and selective science. The Soul as a psychological function is indisputable. Well-meaning people can disagree about the Soul's origins and characteristics. But a decision to deny the Soul's existence leads to an unwise, unhappy, meaningless half-life.

Don't be discouraged by doubts you may have about the Soul's existence or your ability to connect with your Soul. Habits and ingrained attitudes take time to overcome, but a connection with the Soul is achievable, and it's never too late. **You can access the same timeless truth that inspired Aristotle, Thomas Aquinas, Gandhi, Mother Teresa, and less famous folk. The greatness in them lives in you to be discovered.**

The time is right
The New Reality requires wisdom. No amount of intellect, education, or knowledge will make you wise. Wisdom exists only with a Soul connection through the INME. The Soul always endeavors to reach us and never abandons us once heard. When you silence your Ego, the Soul and the INME fill that void, and the Ego loses its dominance over your consciousness. It can happen very fast.

You connect with the Soul already
You already have Soul experiences, even if you don't recognize them: "aha!" moments, inspiration, innovation, wisdom, quotes, dreams, imagination, creativity, empathy, and awe. You'll learn to use these experiences and others to connect with the Soul.

FUNCTIONS

Compensating opposition to the Ego

The Soul functions in opposition to the Ego, each providing half of what is needed for a meaningful, joyful life, which the INME unites to achieve our potential and form the capacity for wisdom. (Refer back to the "Two Worlds" chart earlier in this chapter for additional clarification.)

"Knowing" beyond knowledge

The Soul is the source of information unavailable elsewhere: timeless truths, natural laws, patterns, archetypes, values, inspiration, creativity, and more. But as explained further in the next chapter, "Cultivating Consciousness," it is not only the information that is unique, it is the process by which we receive it. It is instantaneous "knowing," experiential learning so powerful that we are transformed in small or life-altering ways.

Perspective

Connecting with the Soul can offer a momentary glimpse into timelessness or a different kind of time. In those brief encounters, intimacy with all things is felt as you know you are part of the undifferentiated, humble, magnificent reality of the macrocosm-microcosm.

Aspiration

You sense and share the Soul's aspiration for your potential, humanity, Earth, and creation. It is wholly unlike ambition. It is like watching your children or grandchildren play. You want the best for them but understand that their lives will play out with you increasingly being a spectator, each year in cheaper seats farther up from the action. We have in those moments perhaps something like the Soul's aspiration, a loving, patient form of intense, hands-off hope. The Soul is an aspiration for intimacy, empathy, and love with all in creation. And at the Soul's essence there is an aspiration to innovate and create. **The Soul's aspirational effect on us is assurance that creation, whatever its source, is benevolent, loving, timeless, and infinite.**

Continuity

No one knows for sure what happens to the Soul after our bodies die. It may contain an ongoing record of one or successive temporal existences, like how DNA guides biological evolution. If true, our physical existence and life experience would evolve in parallel. Support for this theory is the law of energy conservation, which says we change into a different form of energy at death. Anyone who has witnessed death experiences "something" leaving the body. Could it be energy encoded with the person's life experience? An argument for that possibility is that humans have always designed automation in our image, often unconsciously. Today, anyone can digitize their personality and life experiences into an artificial intelligence digital twin to teach and interact with future generations. This could be a human projection of the Soul's continuity function.

THE SOUL'S VOICE

The Soul has an unmistakable voice or quality to its messages, which is unsurprisingly opposite of the Ego. The Ego never shuts up and yammers on and on, lying and deceiving you to ensure immediate survival and safety. The Soul's messages have a pristine originality, clarity, and brevity. They offer unmistakable meaning, emotional and physical energy, and timeless, undeniable truth that is otherwise unavailable. Many quotes are in fact the Soul speaking through someone with a message to all of us. Our reactions to a Soul message may be a gasp, ecstasy, an increase in energy, chills, intense gratitude, awe, or humility.

MULTIPLE WAYS TO REACH US

The Soul speaks to us in many ways. Our task is to learn to recognize, capture, and understand the implications of every word.

1. *Unexpected breakthroughs*: aha moments, inspirations, creativity, intuitive leaps, ideas.
2. *Artistic expression*: music, drama, dance, painting, sculpting, crafts, hobbies, etc.
3. *Discovery through exploration and sciences*: mathematics, physics, biology, chemistry, etc.
4. *Unifying experiences*: activities that unify body and Soul into one, such as running, exercise, yoga, walking, swimming, sexual intimacy, and losing ourselves in awe of nature.
5. *Kinship, intimacy, and relationships that enable us to experience the shared creative source:* parents of infants often feel it with newborns fresh from the Soul's dimension. Others feel it with animals and nature. We experience it through intimacy with loved ones in those irreplaceable moments where we cannot imagine life better.
6. *Through others*: the Soul often speaks truths or insights through others, even those we would not think capable of such wisdom. Never dismiss the words of children or people society would judge as incapable of wisdom. People who live simply sometimes see us and life clearly when more "sophisticated" people do not. Time-worn truths such as "From the mouths of babes" or someone having "an old soul" apply. Some of the wisest people I have ever known were those whom most would walk by and never notice.

LOCUS, TIMING, STYLE

The Soul's locus is often through "gut feel" or the solar plexus. This may seem strange at first if your Ego has dominated your consciousness, but it is worth attention. Your gut is an alternative source of evaluating life and making decisions that you already know how to use. The Soul's timing for wisdom is not on demand, but when the student is ready, the teacher appears. To make the student ready, we must be receptive and patient.

The Soul's style never judges or criticizes; and is always intimate and unconditionally loving. The Soul does not dictate how you live or believe; its gift is free will through your INME to live as you choose. It is not jealous or possessive and intends us to live a whole, joyful temporal existence in concert with the Soul.

INME

"We are not what happened to us; we
are what we wish to become."

—CARL JUNG

INME DEFINED

Even if you have been exposed to the Ego, Shadow, and Soul, you are unlikely to have experience with the "self" or INME or the process to help it emerge. Yet the INME is the most valuable psychological function because it is our gateway to potential and wisdom.

RECOGNIZING THE INME

Have you heard the voices and thoughts of the Ego, Shadow, Soul, or Conscience and realized that, oddly, they are speakers, and "you" are the listener? Have you agonized over a difficult decision that logic, such as pros/cons and other analytical approaches, couldn't resolve, but suddenly, a wise resolution brilliantly appeared? Who arbitrated the solution? Where did the wisdom come from?

That is the INME. You cannot dialogue with your INME as you can with the Ego, Shadow, Soul, and Conscience. You sense or feel it instead. You feel genuine, collected, and comfortable in your own skin, with the best of your thoughts, feelings, and actions. Your heart, where most people experience their INME, feels like home. The heart is the nucleus of the psychological body, conveniently located halfway between the brain as the locus of the Ego and the gut or sternum, the Soul's portal.

FUNCTION AND EXPERIENCE

The INME is the unique conscious "you," your destiny and reason for living. It is a fully conscious, self-realized, and actualized human operating at its potential, unifying temporal, and eternal realities.

THE INME INFLUENCES THE EGO, SHADOW, SOUL, AND CONSCIENCE

The INME mediates and coordinates the other psychological functions, like a computer operating system. Interaction with the INME can alter the other functions. The Ego can change from volatile to stable, nagging to advising, and from resisting going quiet to compliant when unneeded. The Shadow can become a helpful sixth sense instead of a snake coiled to strike. The Soul speaks directly and frequently to the INME instead of occasionally and subtly. As the INME lives consciously and purposely, interactions with the Conscience become unnecessary and rare. All the functions cooperate to help you see the reality of the personal and collective, and temporal and eternal worlds.

THE INME CHANGES WHO YOU BECOME

The INME expands your capacity for wisdom while remaining dynamic and worldly. You will be humble but willing to contribute your talents. The INME energizes you to be joyful, exuberant, grateful, and optimistic while accepting the reality of life's rhythm, polarity, and adversity. Your patience with yourself, others, and the world will replace perfectionism and judgment. Your growth will accelerate as you seek clarity and understanding about life. **You cease being fearful or anxious as the eternal becomes a reality instead of a concept, and temporal completeness becomes a transition to eternity.** You will experience intimacy, unconditional love, and empathy at a depth rarely grasped and enjoy it in more meaningful relationships. In time, you will establish intimacy with all persons and all things.

INME PROXIES

Third parties can be proxies for your INME while yours develops. Wise parents, grandparents, friends, mentors, or therapists can be temporary stand-ins for your INME. They must be people you respect for their wisdom, trust to have your best interests at heart, and can assist you without telling you what to do. They can be people you know personally or those you know through their works, such as books or videos. Remember, these teachers appear when you, the student, are ready to learn.

ARRIVAL AT YOUR POTENTIAL

The INME is fully established when you are "hitting on all cylinders" or "in the zone," using all your talent and energy in the temporal world through your Ego/Shadow in perfect tandem with your Soul/INME guiding your ethics, purpose, and connection to the collective. The INME is heaven on earth, the art of living well.

Use this QR code to learn more about the INME.

PRACTICES TO USE THE PSYCHOLOGICAL BODY

Following are practices to discover your INME; to get to know your real self and your psychological body substantially better.

PROJECT 1:
DISCOVER YOUR BASELINE

The objective is to discover and know your reality as it is without judgment, regret, or overly focusing on why. There are three practices to gather this perspective that you may choose to repeat several times or throughout life.

First practice: discover me before age six

Defining childhood personality traits clarifies our unique talents and strengths. Knowing them is invaluable. They will be with you for a lifetime. They can be honed and matched to our life's work and avocations. Those same strengths, in excess, create our most significant vulnerabilities, which I call crucibles. Unless we are aware of that risk and act to moderate excesses, they will haunt us repeatedly throughout life, presenting threats to health and happiness, even survival.

Meet with everyone who has known you well from birth: grandparents, parents, siblings, other family, and friends. Ask them to describe you as a child in as much detail as possible, focusing on dominant talents and traits, including those admirable and less so. You will need to encourage them to be candid because you seek a realistic self-portrait instead of a complementary fantasy. Ask them to identify those childhood traits still evident in you today. Ask them to describe traits, talents, or less admirable qualities they see in you today that they never observed or anticipated in your childhood. Request insights they might have into how the unimagined traits came about.

Put on your best poker face and let the people you interview do the talking. Record the interviews or keep detailed notes. Consolidate the interviews into the most thorough description possible. You will find that information extraordinarily valuable immediately and throughout your life.

Second practice: personality predispositions and "gut feel" decision-making

Everyone occasionally has "gut feelings." The saying "trust your gut" expresses homespun confidence that our gut (Soul) gets things right that our heads (Egos) may not. "Gut feel" is different from emotions. It is an intuitive decision-making capability that is the opposite of thinking or logic, like a sixth sense to read intentions or "read a room." The *Oxford English Dictionary* defines *gut feel* as "an intuitive faculty giving awareness inexplicable in terms of normal perception, as in "Some sixth sense told him he was not alone.""[4]

Thinking (logic) and feeling (sensory intuition) as opposing decision-making predispositions came from Carl Jung's groundbreaking work on personality types over a century ago. Jung's personality typology has become familiar to tens of thousands of organizations and millions of people worldwide through the Myers-Briggs Personality Inventory. It played an essential role in my personal development and in assembling collaborative, high-performing leadership teams with complementary strengths.

Complete the Myers-Briggs personality inventory to gain insight into natural diversity, individual differences, and personality predispositions, especially how you make decisions. People are predisposed to making decisions using gut feeling or thinking, a continuum similar to a bell curve. The more adept you are at both types of decision-making, the better able you will be to call on both and make wiser choices.

Third practice: me today

1. **As Others See Me**
    You are now armed with a portrait of yourself as a child and an objective diagnostic from the Myers-Briggs Personality Inventory. Reach out to another panel of people you trust to be candid. Ideally, these are your present-day contemporaries, friends, co-workers, family, bosses, or mentors. Ask them to describe

traits and behaviors they most admire about you and what they wish was different or moderated. Document what you learn in a "Me Today as Others See Me" document.

2. **As I See Myself: Who Am I? (WAI)**

    Every year for decades, I conduct an annual self-assessment. It has been essential to my growth and the emergence of my INME. Take several hours over a week or so. Thoroughly assess yourself and your life now and where it is going. Be as candid as possible. Consciously engage the INME by avoiding self-judgment, which comes from the Ego. Focus on accepting where you are and setting a few goals for self-improvement. Start modestly and become more aspirational with experience. Document your assessment and look at it at least every quarter during the year. When you perform the WAI process each year, look back on prior assessments to gauge your growth.

PROJECT 2:
QUIETING THE EGO

If you are already successful in calming your Ego and connecting with the Soul, you may not need to do much other than do it more often to accelerate your development. If you are unable to calm your Ego, try the following practices. You will need to be patient and try them several times.

1. Use the technique you use today to calm yourself when distressed or upset, to collect yourself, heal, or contemplate.
2. Seek intimate moments with loved ones. While in that loving place, retreat inside yourself. You are seeking freedom from your Ego in order to commune with your Soul. Toward that end, take some time by yourself to say to your Soul, "I am ready." Stay calm and patient, listening and paying attention to feelings in your sternum or gut.

3. Engage a particular practice or place where you connect with nature and feel moved by it. That might be your garden, a favorite walking route, a sunset or sunrise, or whatever moves you. With the Ego silenced, mentally signal your readiness to the Soul.

PROJECT 3:
CONVERSATIONS WITH PSYCHOLOGICAL FUNCTIONS

Practice identifying the Ego, Shadow, Soul, Conscience, and INME by their distinct feel, voice, and thoughts. It will take practice. Focus on one function for a while. Reread the previous sections on each.

Reaching your potential and becoming wise requires active, purposeful engagement with your entire psychological body. You can do this. Give it a try, have patience, and trust the inner guidance system to help you along if you do your part.

WISDOM STORY:
THE STILL, SMALL VOICE

Anne Ryder has been a friend and source of wisdom for over twenty-five years. She is an Emmy Award–winning reporter, news anchor, and documentarian who is now sharing her gifts with broadcast journalism students at Indiana University and with you through this story. Anne's life and this story reflect her dedication to looking for and listening to the Soul's guidance and applying it through her true self in her life and work.

> In April 1996, I sat with Mother Teresa at her Kolkata, India, headquarters, her eyes shining like bright lights from within. She held my right hand and touched each finger for the five

transcendent qualities of service to others, an unforgettable wisdom master class.

That trip was born of frustration as a television news anchor sick of "If it bleeds, it leads" news. Instead, I sought inspiring interviews like the one with Mother Teresa. She promptly denied it but invited me to come "share her works of love." Surprisingly, without asking, she eventually granted me the last on-camera interview she would ever give.

The time with Mother Teresa was one of many stories of hope, faith, and resilience I recorded over twenty-five years. Some subjects were famous, but many were quiet giants anonymously performing acts of service out of love, connection, and devotion to their fellow humans. I discovered in them common qualities that became my wisdom "top ten," which I now share with you.

1. **Trust the still, small voice.** Create space for it by silencing the Ego's petty demands and trusting where it takes you, such as when frustration led me to Mother Teresa and the still, small voice of wisdom that provides direction through inspiration, impulses, or convictions to act.

2. **Draw your circle small.** Many people do extraordinary things but learn too late that family and friends who love them most are starved for their love and attention.

3. **Be grateful always.** Say thank you in the darkest moments, confident you will emerge stronger and wiser.

4. **Don't judge.** You don't know what others have endured. Create healthy boundaries but without judgment.

5. **Honor small acts of kindness.** They move mountains and melt divisions, creating a force of goodness that ripples outward and replicates.

6. **Your suffering alerts you to others in need.** It is like a pain detector connecting you to others that you may touch briefly but profoundly.

7. **Abandon the illusion of control.** We control very little other than our response to what comes next.

8. **Embrace discomfort.** Uncertainty and being ill at ease can expand perspective and worldview and spur self-discovery.

9. **Cultivate silence.** Prioritize clearing your mind and listening for the still, small voice prompting you to step outside your comfort zone.

10. **Kolkata is where you find it.** Mother Teresa lived among the world's worst poverty but said the greater poverty is loneliness, isolation, and materialism, which can make the rich the poorest.

PART II

Advanced Skills

CHAPTER 5

Cultivating Consciousness

"No problem can be solved from the same level of consciousness that created it."

—ALBERT EINSTEIN

WHAT IF BECOMING WISER isn't about cramming more knowledge into your mind but instead about opening an existing channel inside you that connects to timeless wisdom?

Only humans, so far as we know, have a Soul function and consciousness. Consciousness is the process and experience of connecting with the Soul and an indicator of capacity for wisdom. It is a unique life experience unmatched in joy, learning, and intimacy. The Soul is the source of your innate drive to live to your potential. Consciousness is that turbocharged moment when that growth occurs. Sometimes, the experience is life changing. Other times, it is incremental, like an inspiration, insight, or idea that builds to something significant. The recipient of the growth is the real you/INME, not your Ego. Without consciousness, you cannot be human, find meaning, or become wise.

In Chapter 4, we discussed five components of the psychological body: the Ego, Soul, Shadow, Conscience, and INME. The psychological body is how we experience existence. Consciousness, which we will explore next, is how our experience of existence expands to enable us to comprehend the greater reality. Consciousness is cultivated from the time you experience the Soul until your INME absorbs its learning and applies it in the world.

OVERVIEW

Consider these three essential questions.

WHAT IS CONSCIOUSNESS?

In short, consciousness is a particular type of instantaneous experiential learning that substantially differs from traditional learning.

WHY SHOULD I ACTIVELY CULTIVATE CONSCIOUSNESS?

What we discover through consciousness cannot be learned by other methods. Without consciousness, we live instinctive lives, unable to understand reality, meaning, or the opportunity to become our best selves to live wisely and well.

HOW DO I CULTIVATE AND MAXIMIZE CONSCIOUSNESS?

Give it the priority it deserves, as if your life depends on it—because it does. Be open and curious. Have a clear, committed intention to become your best self, to your benefit and the benefit of others.

CONSCIOUSNESS AND TEMPORAL LEARNING CONTRASTED

For practical purposes and as a helpful metaphor, consciousness can be considered a way to learn, like traditional education, except from a different source, for a different purpose, following a considerably different process.

We are surrounded by learning opportunities throughout life: experts or people we admire in classrooms, hands-on experiences, or mimicking their examples, from reading, listening, or watching books, podcasts, training videos, documentary films, movies, theater, etc. In all these cases, other people, living or past, interpret their knowledge, experience, and reality as they see it for us. Some scholars, philosophers, scientists, and futurists go beyond knowledge and experience to speculate about timeless truths, laws, and reality beyond our five senses and logic. They make intuitive leaps that others do not see. We also learn by direct experience on our own without others' interpretations. When we avail ourselves of these opportunities to expand and deepen our understanding, our knowledge and earthly skills expand to the degree we retain what is offered. If we don't use the knowledge, we will likely forget it.

Consciousness is much more focused. The Soul is the single source and teacher. It knows everything about us because the function is inside and outside; thus, it is the ultimate embodiment of personalized learning. The single purpose of Consciousness is to help us become our best and reach our potential. As temporal learning tries to explain earthly reality, the Soul explains infinity's reality. Where earthly teachers look at our exterior and guess what is inside, the Soul looks at our interior and imagines what we are capable of in the temporal world. Temporal learning is a tedious, complex process of capturing through our five senses, sorting, analyzing, and prioritizing. Consciousness is instantaneous experiential knowing. By combining temporal learning and learning through consciousness, we become whole, living comfortably in both temporal and infinite worlds, giving our best as the wise INME and passing it on to others.

STAGES OF CONSCIOUSNESS

Consciousness has three successive stages, usually occurring in a few seconds or minutes but sometimes longer.

1. **Transcendence is when we experience the Soul's distinct, clear, concise voice, often accompanied by physical manifestations such as a gasp, chills, pounding heart, energy burst, exhilaration, or a level of intimacy that can almost be overwhelming.** The *Oxford English Dictionary* defines the experience of transcendence as "beyond or above the range of normal or merely physical human experience or surpassing the ordinary; exceptional."[1] Virtually everyone has transcendent experiences, even if they do not recognize them as such at the time. They include aha! moments, Soul connections in a meditative state, a state of awe connected to nature, or significant events such as moments of deep friendship or childbirth.

   Some transcendent experiences change us forever. Others are less powerful and complete, such as inspirations whose purpose is unknown. It is best to treat everything from the Soul as a mosaic piece that may coalesce into something of significance. Experiences that seem opaque or incomplete should be recorded, or they may fade.

2. **Knowing is the instant when the Soul's special knowledge expands our previous level of consciousness.** Instant knowing seems impossible, but it is genuine, like learning (once) not to pick up a hot pan barehanded or that magic moment when you know someone loves you. Only transcendent experiences provide instant "knowing" that expands consciousness.

3. **Upgrade is when the real you/INME is permanently changed due to the unique knowledge and Soul experience.** The change can be slight or dramatic, but we are a new version of

ourselves that knows everything we did before, what we now know, and how to apply it in the temporal world. The temporal parallel is a smartphone operating system upgrade, except that Soul upgrades never have bugs.

DIFFERENCES FROM TRADITIONAL LEARNING

Learning pace

Traditional learning is slow. In group settings, the pace is shaped by those who are least prepared or slowest to grasp the information. We must schedule our time to match when learning opportunities are available Consciousness is personalized, experiential, and instantaneous. One minute, we don't know; the next, we do.

Pedagogy

Traditional learning requires our five senses. Consciousness is experiential learning. If you have had transcendent experiences, even consequential aha! moments, you know what that means. If you are skeptical, consider technology futurists like Ray Kurzweil, who predicts that direct, noninvasive, user-controlled electronic connections between our brains and automation (brain-to-computer interfaces or BCIs) will become commonplace within twenty years.[2] Virtually instantaneous experiential learning and shared experiences with others who are similarly equipped will become possible. Elon Musk's company Neuralink has developed an implantable version of such a BCI to treat musculoskeletal diseases. The Food and Drug Administration has approved clinical trials of the device.[3] There are dozens of other companies like Neuralink in various stages of development. This doesn't suggest that technologically enabled experiential learning replaces consciousness, because that requires the Soul. It does suggest traditional learning could occur similar to the speed of consciousness, a staggering thought.

Readiness to learn

Traditional learning depends on the teacher, the education system or employer, and us to determine our readiness to learn. We use qualifiers like age, maturity, previous education, and tests to assess readiness. The error rate is high. Soul learning is always accurate, and the Soul is always on duty. Because the Soul is both outside us in the macrocosm and inside us through our Soul function, it knows precisely what, when, where, and how we are ready to learn, comprehend, and put what we learned to use. This is the meaning of the saying, "When the student is ready, the teacher appears."

Brevity and retention

We take in a lot but retain relatively little in traditional learning. Curricula developers, teachers, and trainers cannot know what we know or need, so they give us volumes of material and hope that repetition will reinforce their lessons. Contrast this with how we learn through the Soul. Its communication is distinctively brief, perfect, and rich in meaning. A few words inspire complex thoughts, requiring thousands of words to explain. Consciousness represents a perfectly digestible amount; we forget almost nothing and apply it all.

Learning sought and learning delivered

For most of our lives in traditional learning, we seek the teacher rather than the teacher showing up unannounced and uninvited with a lesson. The Soul's teachings do both. We seek the Soul when we meditate or pursue an activity or psychological state where our Ego is quiet, and we hope the Soul will speak. It happens only if our intention is intimacy with the Soul. Other times, we go about our day with no thought of connecting to the Soul, but because we are open and perhaps for other reasons known only to the Soul, we receive inspiration, insight, aha! moments, or other unexpected consciousness experiences. The more we take these events seriously and keep a record of them, the more experiences we will have.

THE SOUL'S TEACHING AS METAPHOR

A grandfather takes a daily seaside walk with his four-year-old grand-daughter. Standing ankle-deep in the ocean, he explains the laws of the sea and phenomena that she finds interesting and can grasp, such as why water swirls around their ankles. He encourages the child to write or draw what she learns and make her own observations because the sea and life play by the same rules. Her attention sometimes wanders or is captured by a seagull. The grandfather patiently smiles and waits for her attention to return. Maybe today, maybe tomorrow, perhaps further in the future. But the grandfather's teachings stick. As the granddaughter grows into an adult, she is amazed by how much she remembers. Her lessons with her grandfather continue, and her learnings become more profound, guiding her adult life, decisions, and values. Slowly but surely, the now middle-aged granddaughter's consciousness and wisdom expand. She fondly remembers the value of the lessons and commits to being an equally devoted grandparent. This metaphor conveys how the Soul's teachings work, as well as our responsibility to pass along its wisdom.

WHY ACTIVELY CULTIVATE CONSCIOUSNESS?

OPTIMAL PERFORMANCE

Consciousness plays an essential role in achieving potential and wisdom. Without it, we rely on the Ego to learn and retain information. The best we can become is knowledgeable but soulless, a weak imitation of the person we could have become. Sadly, that is the case with many people.

THE UNIQUE EXPERIENCE

Consciousness produces growth and understanding of our inner and outer worlds in ways that intellectual learning could never achieve. Author and mathematical physicist Roger Penrose describes it this way: "Consciousness is the phenomenon whereby the universe's very existence is made known."[4]

Consciousness can only be understood by experiencing it, but there are approaches short of experience that can help.

- Examine events that you treated lightly at the time but now recognize as transcendent.
- Discuss consciousness and the transcendent experiences with people you trust.
- Discuss consciousness with a trained PhD therapist, particularly those who are growth-focused and have experience assisting people with techniques to calm the Ego.
- Relate the consciousness experience to things you know, such as temporal learning and the grandfather metaphor above.

Something powerful and wonderful awaits us beyond the Ego's dominance and day-to-day activities, consuming time, and attention. I and many others I've known have had life-changing experiences. Consciousness requires patience and determination, but it is one of the miracles that make us human. Collective consciousness is also ascending, and you can learn from it.

INME DEVELOPMENT

Consciousness accelerates INME development. You can suddenly be shocked by your energy level, focus, and clarity. Multiple aspects of life can improve simultaneously, with little or no effort. **A few transcendent minutes or hours over a lifetime is better than a PhD in philosophy**

or thirty years of trial-and-error life experience. You can become wise beyond your years because you experience undiluted timeless wisdom firsthand and know how to return to the fountain from which it flows. You experience a slight increase in progress to your potential reflected in work, finances, health, intimacy, and lifestyle that accelerates in time.

CONSCIOUSNESS SETS OFF A STARBURST

Consciousness expands in every direction like a starburst, broadening and deepening comprehension of many aspects of life that we would not logically connect with the Soul's message at the time. This adds unpredictable dimension, uniqueness, maturity, and depth to your INME.

THE SOUL IS EVERYWHERE

The Soul's voice is all around you, just like the Ego's voice. Once you recognize the Soul's voice by its clarity and wisdom, you can expand your transcendent experiences and consciousness by remaining aware of it and open to its messages. Your antenna goes up to sense the Soul's presence in anyone, anytime, and to see and feel it in nature as never before. The wisdom stories included in each chapter of this book contain the Soul's voice.

Experiencing the Soul's ubiquity changes you forever. It becomes an opportunity instead of an effort to connect with people who momentarily cross your path. You realize that everyone carries the Soul inside them, even if they didn't request it and refuse to acknowledge it. Forgiveness becomes easier and then a natural first response. Each experience is a learning opportunity that enriches your INME.

INCREASED INTIMACY WITH THE SOUL

Once the connection to the Soul is activated and practiced, creativity and inspiration increase. Your confidence in yourself and life becomes unshakable. You worry less and live better because the Soul never demands, only informs. It is always calm and optimistic.

PERSPECTIVE BROADENS, DEEPENS, AND ELEVATES

Connectedness

Consciousness makes you more aware of connectedness. Understanding sets the stage for more transcendent experiences. For example, you can observe how your energy connects to others in a crowd. If you sense unrest, you become agitated and defensive. But if the mood is upbeat and excited, you will be as well. If you share the awe of a sunset, you and the crowd become peaceful. If you show calm when others are rattled, they will look to you for guidance.

Calm courage

You will become increasingly calm as your intimacy with the Soul grows stronger. You will understand yourself and life's reality far better than ever. You will know your place in the world and why it matters. You build relationships with people who bring out your best rather than your worst. Your security comes from self-confidence instead of chasing endless Ego demands such as if I earn X amount, I'll have "enough," or if I meet my soulmate, I'll be "happy," and so on. Instead, you develop calm courage that whatever happens in life, you will be safe as long as you do your best to be your best. These and other examples of the art of living well are only available through the INME's expanding consciousness.

CONSCIOUSNESS PRACTICES

MAKE CONSCIOUSNESS A COMMITTED PRIORITY

The seriousness with which we treat these practices matters. It is how the Soul detects whether you, as the student, are ready. If you are skeptical, half-hearted, or uncommitted, correct it before attempting to connect with the Soul. That counts as progress.

BE CLEAR ABOUT YOUR INTENTION

Unclear or misguided intention is a barrier to connecting with the Soul. Your intention should always be intimacy with the Soul or expanded consciousness instead of something you "think you should do" or to satisfy others. It means openness to insights and truths the Soul offers instead of asking the Soul to fix your life or answer questions you should answer.

OPEN YOUR HEART AND MIND

Your jealous Ego is the only thing between you and the Soul and consciousness. Openness and receptivity is the state you seek. You are hardwired to grow, and your Ego cannot answer life's most unavoidable questions or make wise decisions. When you open to the Soul, consciousness expands.

EMBRACE THE DYNAMISM

Consciousness, like nature, has a rhythm that your Ego can never understand. Maybe someone says something, or you see something that triggers an aha! moment. Or unexpectedly, you have an insight watching a movie, walking, or driving. Recognize its importance and value.

SEEK QUIET CALM

The Soul function speaks on its agenda in a time and place of its choosing, but you can signal your desire and readiness. For decades, I have reserved specific quiet times for receptivity. My business and life success and the Potentialist work came from those moments. I may have interests or things I'd like to learn, but I ask for and expect nothing. The Soul already knows what I need and when I am ready. My best times are the first hours of each morning as I enjoy sunshine and coffee, writing, and journaling and later in the day during a solo workout, hike, or walk. Evenings before falling asleep are also special.

You have likely sought quiet and waited for inspiration in the past when troubled or something baffled you. Maybe you went for a walk or sat on a bench and waited patiently for insight. That's the state of mind and receptivity needed. A mentor of mine described it as a state of patient eagerness. It has also been called active openness or active receptivity. The words appear oxymoronic, but together, they are an expression of "both," your INME's ability to bridge the temporal and eternal, which is impossible for your Ego, so the Soul recognizes the INME.

Meditation is another ancient way to enter a serene, receptive state. Everyone should try meditation, but it doesn't work for some. Don't give up on connecting to the Soul if it doesn't work for you. Hyper-busy lifestyles keep the Ego agitated. Failed attempts to meditate can create frustration, anxiety, or disappointment, creating a barrier to the very peace you seek.

CAPTURE EVERYTHING YOU HEAR
FROM THE SOUL FUNCTION

Record every word and image that comes during transcendence. Assume all is irreplaceably valuable because it is. Capturing what you experience signals that you are paying attention.

SHARE TRANSCENDENT EXPERIENCES
RARELY AND THOUGHTFULLY

The invigorating or even intoxicating effects of transcendent experience may cause you to share your excitement with others. However, doing so can dilute the effect and value of your experience, especially before its meaning is fully digested or when shared with someone who judges it. Keep these experiences to yourself until you can grasp the complete picture and then only disclose them if the person needs to know or can appreciate and benefit from your experience.

EXPECT THE COLLECTIVE EGO
TO OPPOSE CONSCIOUSNESS

As you read this book and observe life through a different lens, you should be able to recognize the Ego and Shadow: yours, others, and the collective. It's easy because the Ego's survival-at-all-costs mission attempts to maintain total domination of consciousness. It can be nasty and dangerous when threatened. Many of history's darkest deeds were perpetrated by fearful, threatened, jealous, possessive Egos and Shadows. Even today, intelligent, seemingly sophisticated people can be the collective Ego's brown-shirted stormtroopers. They respond cynically, disparagingly, or aggressively at the mention of Soul, consciousness, free will, and human potential. Nothing is achieved debating them. That isn't the way of our inner guidance system, the Soul, or consciousness. People must come to the Soul or respond to its offers openly and willingly.

NEVER SPEAK FROM PRIDEFULNESS

Pride in newfound growth, insight, and wisdom is understandable, but it comes from the Ego. Sharing from pridefulness always falls flat; I've seen it many times. Say to yourself, "This is a gift; my potential is my duty to seek," or something similar to return to your INME through humility.

SUMMARY TAKEAWAY

The Soul is always trying to help you reach your potential through consciousness. It speaks to you constantly if you listen and observe. Open yourself to its messages and the rewards of consciousness and wisdom. When you're willing to listen, the course of your life will change. Don't get hung up on what the Soul is or isn't. Simply welcome it as a function of the psychological body and allow it to help you learn the art of living well.

Use this QR code to learn more about consciousness.

WISDOM STORY:
KNOWING WHEN IT ISN'T ABOUT YOU

Pattiann Gavaghan has been a dear friend for over four decades. She's an adopted aunt to my children, a revered elder to my grandchildren and extended family, and my go-to for wise counsel. Her story is instructive.

I was raised in a typical Irish-Catholic family, the fourth of five siblings and the only girl. I was proud of our family and loved being the sole daughter without appreciating the future responsibilities involved. Over fourteen years, I lost my parents and three of my four brothers. It was a tragically difficult time but an incredible learning experience, like all adversity.

My eighty-four-year-old mom died of a heart condition. She couldn't recognize me at the end or tolerate my presence. Heartbroken, I wanted to plead, "Remember me!" but that would only have increased her suffering. Her dying was about her, not me. My job was to make her passing peaceful.

Two years later, my ex–Boston Marathon athlete brother Bill died at fifty-nine from a heart attack while on a leisurely run. It was heartbreaking not to express my love in a goodbye. But I wasn't alone. My dad lost his son; my brothers lost their brother. It wasn't about my grief but enduring mine while helping them through theirs, especially Dad.

Five years on, my brother Jim died of esophageal cancer at sixty-five after a valiant fight. I again wanted to say goodbye and express my love. But Jim's way of dying was refusing to acknowledge it. His death wasn't about my wants, but his hope until the end to beat cancer.

A year later, my dad, a fixture in our community, died at ninety-seven. I was his buddy, event companion, and caregiver during a long, painful decline that was frustrating but full of gifts and laughter. I wanted to capture Dad's memories, good and bad, but he was from the silent generation that didn't share feelings easily and chose to remember only the happy ones. Again, it was his happiness that mattered, not my curiosity.

Only five years later, I was shocked when Dennis, one of my two surviving brothers, was diagnosed with early onset Alzheimer's. I walked on eggshells, learning what to say and do to avoid upsetting topics such as his prognosis. I discovered he loved playing golf with me, which we did every summer until his death at sixty-eight. It wasn't about me reassuring him but letting him forget his fate for a few enjoyable hours.

Each death is the dying person's unique experience. None of it is about us, as much as our Ego would make it that way. It's a desired peaceful transition that everyone needs. During difficult or celebratory times, we frequently make things about ourselves that should be about another's experience. It's an unconscious, selfish behavior worth avoiding.

CHAPTER 6

Conscious Intention

"We either live by intention or exist by default."
—KRISTIN ARMSTRONG

WHAT IF A SLIGHT CHANGE IN SOMETHING you already do unconsciously could significantly improve your career and personal relationships and open doors to your potential and wisdom?

Rarely do we appreciate the central role of intention in career and personal relationships. Most of us have never been taught the importance of communicating our intentions to others. Almost no one is aware of intention's vital role in the psychological body, particularly in how we relate to the functions of Ego, Shadow, Soul, and Conscience.

Intentions can be conscious, healthy, and clear, or unconscious and undefined, leading to misunderstanding or misuse. Intention is one of the arts of living well and a gateway to consciousness, potential, and expanded capacity for wisdom. **Conscious intention is easy and quick to learn because we already practice it occasionally; we only need to make that practice habitual.**

INTENTION DEFINED

"Intention" has many synonyms, including plan, goal, purpose, object, scheme, desire, ambition, dream, hope, aspiration, and calculation. Unfortunately, none capture conscious intention's essential role in potential, wisdom, and life success.

Imagine awakening in a vast desert with no landmarks or water. Already thirsty, you recall a map that identified a water source in the desert's northwest corner. But it's a cloudy night, and you can't orient without the guidance of stars or the sun. Without a compass, you are prone to wander aimlessly and risk dying of thirst.

Intention is a psychological and spiritual compass that keeps us on the path of building character and living well. Intentions that honor us and others support us in developing the wisdom to do the right thing. Honorable intention becomes an ever-present north star. As Deepak Chopra has said, "Intention energizes. Intention transforms." Without water, your energy will be quickly sapped. With a compass, you will find the water source you need and feel reenergized. Like intention, that compass transforms you from a lost victim to a confident, successful traveler.

Every wise person I have known has been described as "guileless," or "an open book." "You always know where they are coming from" is another apt depiction. They are not naïve, but the opposite. They can be stoic, reserved, and thoughtful, but they are not manipulative or vengeful. Unsurprisingly, people sense their intention to do the right thing, the wisest thing. As a result, their relationships are steeped in mutual trust and respect.

How you use intention determines its capacity to harm or help. Wielding conscious intention is like learning to use functional but dangerous tools such as knives. Careless, reckless, or unconscious intention will lead to injury. A responsible adult can safely handle a sharp knife with some training and good habits. Occasionally, we get a careless cut, reminding us to be more careful. Just as our mind effortlessly signals caution when we reach into a drawer for a razor-sharp knife, habitual conscious intention is activated when we consider, choose, and disclose intention. **Used wisely, intention becomes a precision relationship tool.** Habitual conscious intentions improve career and personal relationships and quality of life.

Most of us have had an unpleasant or hurtful experience where our intentions were misunderstood or we were called out for being manipulative or deceptive. We've likely had a situation where our confused or unclear intentions caused trouble or heartache. These problems can go away and never return if we make a little consistent effort.

THE UNAPPRECIATED IMPACT OF CONSCIOUS INTENTION

The importance of intention has been largely ignored and rarely taught because good intentions are the assumed societal norm in much of the world. In his remarkable book *Talking to Strangers*, author Malcolm Gladwell explains that we are wired to assume positive intentions in others. Gladwell's proposition is reinforced by the truth-default theory (TDT) discovered by researcher Tim Levine. He writes that we are unskilled in detecting lies because "We have a default to truth; our operating assumption is that people we are dealing with are honest."[1]

Assumed good intention enables society to function effectively. Otherwise, every daily transaction would begin with suspicion or mistrust, and life would be miserable and unproductive. If you have traveled to parts of the world that operate under the suspicion assumption, you quickly experience its uncomfortable dysfunction. Still, Gladwell recommends some caution. Assuming good intentions in others and expecting them to assume ours exposes us to risks of damaged character, personal brands, relationships, and opportunities.

INTENTION IN THE
TWENTY-FIRST CENTURY

CHALLENGES TO THE SOCIAL NORMS
OF PRESUMED GOOD INTENTION

Accelerating pace of change and life challenges the societal standard of assumed good intentions. Aging societies, electronic communications, and collaborative work across varying distances, cultures, and time zones require thoughtful and sophisticated communication of intentions.

The assumption of good intentions is grounded in our knowledge and confidence of people, neighborhoods, communities, and cultures. We trust what we know. Through technology, we are increasingly connecting with a broader variety of people we don't know. Inevitably, we interact, both online and offline, with people who may not share our values. We encounter scammers and thieves more frequently because worldwide electronic communications have made it possible for them to prey on us through emails, texts, and robocalls. Artificial intelligence, deepfakes, and other technologies will require us to question the authenticity of anything we see, hear, and read. New technology will emerge to help, but caution will unfortunately be part of twenty-first-century electronic communications. All of us have experienced how easily cryptic texts or email messages can be misunderstood, leading to relationship damage.

DELIBERATE DISHONORABLE INTENTIONS

Some individuals, professions, businesses, politicians, and their agents accept or even endorse misleading, manipulative, and undisclosed intentions. You have likely witnessed egregious behavior and abandoned ethics excused by the statement "It's just business (or politics)," as if the participants were gangsters excusing their behavior instead of professionals. This corrupting attitude is as old as humankind, but it is also often glorified in movies, TV, theater, and the media instead of being scorned as in the past.

A corrupting fiction is that someone can be privately honorable but professionally unethical. It is destructive to everyone involved. Such ethical tension splits the personality and makes psychological health impossible. The practitioner knows they are a fraud vulnerable to discovery. Their Ego will drive them to greater extremes, and their Shadow will persistently remind them of their vulnerability. Their fraudulent life becomes a living hell to all but the amoral. This kind of ethically siloed life is more likely in those who equate success with wealth, fame, power, social status, and appearance instead of character and doing their best to be their best.

CAUSE AND EFFECT
IN ELECTRONIC COMMUNICATIONS

We have already seen how people deny the law of cause and effect because they cannot see the impact of their actions and assume that they will not suffer consequences. Electronic communications increase the denial of cause and effect. We all witness people becoming aggressive or nasty on social media or email to a degree that they would never risk in person. When I led companies, I came down hard on email nastiness and more than once marched the perpetrator to the recipient's office and demanded a face-to-face encounter. Without exception, the perpetrator backed down. Electronic aggression is cowardice.

Similarly, people who concoct electronic scams and frauds are often from other countries and view faceless victims as opportunities without consequences. These are fools who will pay without knowing when, where, or the price.

The prevailing social norm of assuming good in others will likely change over the next few decades. Conscious, clear, aligned, and transparent intentions will become an essential twenty-first-century skill. Our challenge will be to preserve the assumption of good intentions while accepting the manipulative power of the digital age. As the Russian proverb says, "Trust, but verify."

INTENTION AND ALIGNED INTERESTS

Collective success in any endeavor requires aligned intentions and interests. Alignment is achieved when intentions and interests are conscious, clear, and declared. Alignment is virtually impossible when intentions and interests are unknown, confused, conflicting, or undisclosed. Even experienced leaders sometimes overlook this essential leadership quality.

Relationships stand the test of time by aligning the parties' intentions and interests and balancing the needs of each involved party. A close study of your life will almost certainly reveal significant time and energy devoted to aligning the intentions and interests of the people and organizations in your life.

Much of corporate leadership is related to aligning intentions and interests among shareholders, customers, suppliers, employers, communities, and government. Democracies and capitalism are systems that align the intentions and interests of political and economic stakeholders. General Dwight David Eisenhower commanded World War II Allied Forces in Europe instead of more experienced combat generals because he was outstanding at aligning intentions and interests to get results. After the war, he was a successful two-term US president practicing his art.

Any leader who has orchestrated collective success in sports, business, or even a family outing appreciates the challenges and power of aligning intentions and interests. Everyone selflessly pulls in the same direction to reach a goal. Misaligned, undisclosed, or self-serving intentions and interests cause disappointing shortfalls. Severe misalignment may damage relationships and careers, social standing, and brand.

RACHEL AND BOB:
A CASE STUDY OF MISALIGNED
AND UNDISCLOSED INTENTION

Bob and Rachel were co-workers and best friends for over a decade. They frequently socialized at the local pub, deepening friendships with each other and co-workers. That camaraderie led to their department becoming a top performer in the company. Rachel and Bob also played in a softball league together for years; their spouses became friends, and their kids grew up together and attended the same schools. Their lives became increasingly intertwined as their friendship prospered.

Unexpectedly, their long-standing, respected departmental vice president announced that he had accepted an out-of-state job to be closer to family. Following company policy, his VP job was posted to attract replacement candidates inside the company, and a recruiting firm was retained to secure outside candidates. Most departmental employees assumed Rachel would be the next VP. She was a top performer respected by the departmental employees and had been at the company longer than anyone else in the department. Rachel had been the VP's right-hand person and backup for years. She wanted the job badly and felt she had earned it. The promotion to VP would mean a significant status promotion and pay increase. She expected the interview process to be a formality due to her track record and tenure.

Rachel was called to the senior VP's office for an interview. She rehearsed her acceptance speech and itemized how to spend the additional VP income. Her expectations were shattered minutes later when she was told that Bob, her best friend, had been named the new VP. Rachel's pain and humiliation were intensified because Bob had failed to mention his intention to apply for the job despite numerous opportunities. She returned to the department offices to confront Bob, overwhelmed by shock, hurt, and anger. Bob explained that he was good at his job and applied because he could be a good VP. His family would benefit from the extra money like Rachel's. Why shouldn't he apply? Bob argued that he and Rachel were equally qualified, and his intentions were honorable and predictable.

His defense only inflamed Rachel's indignation and anger. How could someone who claimed to be a friend see the situation that way? He had betrayed her. She heatedly challenged Bob. At a minimum, she asserted, he owed her a heads-up that he intended to apply. As her anger increased, she demanded that Bob stand down and allow her to be promoted, given how hard she had worked, wanted, and needed the job. Bob adamantly refused. Overwhelmed by her emotions, Rachel hurled vile language at Bob and stalked out of the office.

Unwilling to report to Bob, Rachel quit her job of over two decades and never spoke to Bob or his family again. Rachel and Bob's dispute spread to their co-workers, leading to a departmental rift. Some co-workers believed that Bob had wronged Rachel. Others felt Bob had done nothing inappropriate and that Rachel was being unrealistic and unfair. Years of camaraderie that boosted the department's performance evaporated in a single afternoon.

Bob resigned a year later, unable to win the loyalty or respect of his staff after assuming the VP role. People in touch with Bob reported that his career and life never recovered from losing his long friendship with Rachel, her family, and the respect of co-workers.

Two jobs, lives, and families were damaged, and company performance was degraded at a critical time. This highly destructive situation occurred because intentions were inadequately considered, identified, and communicated. I've witnessed situations like this repeatedly in my life and career. Sometimes, the offending person believed they were cleverly deceiving others by masking their intentions. In other cases, they made an unfortunate assumption that mutual understanding existed only to discover the opposite. Whatever the circumstances, offenders are rarely forgiven and are considered untrustworthy by those who feel "burned" by their manipulative and undisclosed intentions.

INTENTION IN INME DEVELOPMENT

The intention to listen to and act on the Soul's messages germinates the seed of consciousness and the INME. Each conscious intention triggers a growth cycle toward your potential and capacity for wisdom. The Ego embraces anything that soothes its insecurity and increases safety, including confusing, manipulative, and deceptive intentions. The Soul and Conscience urge us to do the right thing (conscious, clear, disclosed intention), and the Ego pulls in the opposite direction. This tension causes the INME to mediate and resolve the opposites through creative resolution. The INME will increasingly urge you to choose honorable intentions as it grows stronger, wiser, and more conscious.

INTENTION AND ACTIONS IN CHARACTER, BRAND, AND RELATIONSHIPS

Trusted relationships and character are generally accepted measures of a successful life. Few would include intention, but it is the essential precedent. **Character is attributed to people with honorable intentions and actions that match them.**

"The road to hell is paved with good intentions" is attributed to Abbot Saint Bernard of Clairvaux (1090–1153). Whatever its origins, this ancient maxim captures the essential coupling of actions with intentions. Good results rarely come from dishonorable intentions. But good intentions do not guarantee that things will work out as intended. People who repeatedly claim good intentions but whose actions are inconsistent are seen as insincere, phonies, or untrustworthy fools. How do wise people bridge good intentions and actions? The most effective method is consistently doing and being your best, including taking actions that reflect your best efforts. No one can reasonably ask more of you than your best. **Most people will assign good character to honorable intentions and best effort.**

You may question how you can be sure you are giving your best or that your best will be good enough in specific circumstances. Perfection is an Ego illusion and cannot be your standard for "best." How others judge you isn't totally reliable, either, even if it's your boss or someone else you feel compelled to satisfy. It is wise to explain risks related to intended results, including your own limitations, unknowns, and unpredictable variables. Sometimes, other people will not accept your best effort as enough. Some are caught in the Ego's distortion of perfection. Others use it to gain power over others, demanding a performance standard that they could not live up to. It is what you know rather than what they think that is important. Give them the benefit of the doubt. Consider what you may have overlooked or other actions you might have taken. Learn as much as possible to improve but trust your gut above all else; it is the only reliable measure. With practice, you will know when you have done your best to be your best. You will feel it, see it, and positive effects and peace of mind will follow. Trust that feeling without second-guessing.

Conscious, honorable intentions lead to accountability for results. Unconscious intentions can cause us to blame others or external influences like fate. If results do not live up to our honorable intentions, learn and take greater care the next time. This "intention-accountability-best effort" learning dynamic becomes a potent growth skill. Parents and leaders play an essential role in developing this skill in children and adults by setting expectations that challenge and encourage growth while respecting and honoring someone giving their best. My most influential mentors were masters in bringing out my best. I hope to have similarly served the people I led in business and my children and grandchildren.

INTENTION'S SUBTLETIES

THE GREETING RITUAL

Pay attention to your gut feelings in the first few seconds of meeting someone new. You will discover that you are observing, listening, and dissecting their intentions. The process may continue for several minutes until you have "sized them up." You may walk away thinking, *I'm not sure why, but I have a bad feeling about that person.* Or you might think, *That person seems trustworthy and genuine,* or *I'm drawn to that person, and they seemed interested in me as well.*

Animals have rituals to detect and signal intention. Horses nibble each other's withers. Dogs sniff bottoms. Giraffes rub their long necks. Animals may be only testing for friends or foes, but more likely, their intention rituals are highly complex. Humans use a "sixth sense" or "gut feeling" to detect and examine intentions far beyond friend or foe. If you listen to your gut and reflect on what you felt after first meeting someone, you may be pleasantly shocked by the intricacy of assessments you performed in seconds or minutes. Since you instinctively know how to analyze your intentions and those of others, why not use that skill to accelerate your growth and improve your relationships?

CHARITABLE INTENTION SHOULD BE SIMPLE; IT'S NOT

In her book *The Soul of Money*, author and humanitarian Lynne Twist explains that some people do good deeds to serve their Ego and reputation rather than to express sincere concern for people or causes.[2] Their true intentions are usually glaringly evident. They resent it if others judge their uncharitable charity harshly. They fail to understand that the honorable intention to do good is damaged or undone in a bargain for recognition.

Honorable intention cannot arise from the Ego's thirst for validation, safety, or self-promotion. Giving with pure intention comes from the

INME as it balances the Soul's call for charity with temporal demands for frugality. Giving freely and charitably sounds easy, but close attention to others and ourselves often reveals the Ego's need for applause. Spanish philosopher, jurist, and physician Maimonides created a hierarchy of gifting intentions that's as relevant today as it was a millennium ago. The giving scale illustrates intention's subtle but powerful role and how charitable giving can become a difficult decision requiring wisdom. Maimonides ranked the least pure to the purest intention as follows:

1.   Donations given grudgingly (least pure intention).
2.   Giving less than one should give but cheerfully.
3.   Giving without a request.
4.   A recipient knows the donor, but the donor does not know the recipient.
5.   The donor knows the recipient, but the recipient does not know the donor.
6.   Both giver and recipient are unknown to each other.
7.   Offering help before it is needed so that the assistance is a leg up rather than charity (purest intention).[3]

Gladwell's, Twist's, and Maimonides's advice demonstrate conscious, healthy intentions' subtle but powerful effects. They also raise another of life's unavoidable questions: *How can I be sure that my intentions are genuine, and my actions are consistent with my intentions?*

INTENTION CASE STUDY: MARY AND HER NEW NEIGHBORS

Mary lived in a nice neighborhood. She was determined to be involved with her neighbors to keep it that way. When new neighbors moved in next door, she delivered a bottle of wine and French pastries as a welcome gift. What could go wrong with an act of kindness? Mary could have paused to clarify her intentions and assess how the new neighbors might interpret them, but she never heard of conscious intention. She rarely asks herself why she's doing something or how others might interpret her actions.

If Mary had quickly reflected, she would have discovered her other intentions beyond kindness. She hoped to make friends with her new neighbors, and her gift seemed an excellent way to begin a friendship. She expected praise from new and old neighbors in return for her kindness. As you might have guessed, Mary suffers from low self-esteem, and pleasing others makes her feel better about herself. But her intentions are more self-serving. Mary's house shares a lot line with the new neighbors; she needs their financial contribution to replace an aging fence separating their back lawns. A housewarming gift could strengthen her negotiating position. Mary's unselfish act was anything but, and her true intentions were unconscious, unclear, and undisclosed. Mary made the mistake of assuming that the neighbors would not see her true intentions.

Mary's housewarming gift could have been well-received if intentions were conscious and transparent, even her desire to be friends or cooperate on a fence repair. But she damaged her character by pretending to be gracious and generous when she was hiding her self-serving intentions. Mary's new neighbors were sophisticated people with a history unknown to her. Most of us are good at sensing ill or manipulative intentions; theirs was well-honed. A single discussion set off their alarm bells. They had a prior negative experience with a homeowner's association (hardly unusual) and self-serving neighbors bearing gifts. They checked with other neighbors and learned that Mary had a reputation for less-than-candid intentions. Her gift backfired, and her new neighbors held her at arm's length afterwards.

INTENTION PRACTICES

Hopefully, you fully grasp the importance of conscious, aligned, clear, and transparent intention and its subtleties. Let's turn to how to make honorable intention an effortless habit using these four criteria: conscious, aligned, clear, and transparent.

MAKING INTENTIONS CONSCIOUS

Conscious intention requires a series of choices. We choose to discover our intention and be clear about it to ourselves and others. We choose an intention that equally serves us, others, and life over a self-serving intention at others' expense. We choose to disclose our intent to avoid being misunderstood or being secretive, manipulative, or insincere. These choices raise intention from shadowy depths and confusion to decisive, overt actions that define our character, relationships, and life.

Choices based on conscious intention take only a few seconds of focused awareness. With practice, you'll be amazed at your speed and precision. It may seem impossible or clumsy at first but recall the complexity and speed of human and animal greetings. Ask yourself: What do I intend in this transaction or situation? Keep in mind that there is always an intention to be identified. Only you know it and own it. In almost every case, our intentions are already evident to us, just below conscious awareness.

We might not like what we discover at first. Our intentions may not be as pure or straightforward as we assumed. Try not to judge yourself too harshly. Intentions can change quickly in the light of consciousness. Sometimes, intention involves complicated, difficult choices that need time to be weighed and examined. Writing down and then reading options and decisions aloud can be illuminating.

ALIGNING INTENTIONS

To discover whether intentions are aligned, ask yourself: *Does my intention align my interests with those of others involved? Does it add or detract from our relationship? From life?* You will likely grasp your own interests quickly. Aligning others' interests and resolving conflicts between your interests and theirs is more challenging. The easiest way to clarify and align others' intentions and interests is to ask them.

Questions you might ask include:

- "Would you outline your goals so I can be sure I'm providing what you need?"
- "What would you like to accomplish?"
- "What is important to you?"

This approach is usually appreciated, and when it's not, that may signal that you're interacting with someone who has impure intention.

Aligning multiple interests is often challenging and imperfect. The more people involved, the more difficult it is, as anyone who has attempted to get four couples to agree on a restaurant knows. However, aligning interests without control is a vitally important skill of being human and a requirement of successful leadership. We become more sensitive to others' interests as we become more conscious and wiser.

You may initially be baffled by what it means to serve the interests of life because we tend to minimize our importance. Remember the unavoidable question that lies at the heart of that, *Does my life and how I live it matter?*

Your every action alters infinity, beginning with your choice of intention. Your Soul knows how to serve life. Guiding you on that subject is its primary psychological function. Listen and follow it.

CLARIFYING INTENTIONS

Clarity is a final communication preparedness check to avoid confusion around creating good intentions and aligning them with others. A helpful metaphor is calling your shot when playing pool. Ask yourself, "How will I explain my intention and interests as simply as possible?" A seconds-long practice conversation in our head suffices for most matters. For more complicated issues, write it down and read it back aloud. Written intentions, alignment, and actions are required when a large group is involved, such as a project at work, in the family, or the community.

MAKING INTENTIONS TRANSPARENT

Once your intention is conscious, aligned, and clear, determine how to disclose it, even if it should be obvious. Transparency of intention is almost always valued and rewarded. People may compliment you with statements like "I always know where you're coming from" or "your candor is appreciated." Declaring intention in a natural, constructive way requires practice and skill. Experiment with techniques to communicate intentions that fit your communication style.

MAKING CONSCIOUS INTENTION HABITUAL

Conscious intentions can become a valued habit very quickly. Reflect on clarity of interests and aligned intention, perhaps while driving home after work or before sleep. When, where, and with whom did you fail to perform a lightning-fast intention check? How did those interactions and situations work out? Can you detect anxiety when intentions are unclear or unknown? Conscious intention simplifies life. As you become more aware of intention in yourself and others, you will become more proficient in noticing how well intentions meet the four criteria outlined in this section; skill increases as we become wiser. Experience teaches us about the people involved and how they behave under stress.

CONSCIOUS INTENTION REWARDS

The rewards of conscious intention far exceed the effort to acquire the habit. The two case studies below and a sample of the many benefits illustrate the value.

AVOIDING CHARACTER, BRAND, AND RELATIONSHIP CATASTROPHES

One painful situation can remind us of the price of confused intentions or interests: a seriously damaged relationship, brand, or reputation; a blown business deal; an angry customer, partner, or friend; or a disappointed boss and mentor. A small effort requiring a few seconds or moments of consciousness could have prevented it.

ASSESSING THOSE WE CHOOSE FOR RELATIONSHIPS

Willingness to be conscious, clear, aligned, and transparent with intentions indicates a person's capacity for intimacy and their level of fear and anxiety. Determining intention requires patience, but it is always revealed in time. Misaligned intent and interests do not necessarily mean the other person is manipulative. They may not understand the importance of intention in relationships and transactions. They may have been taught to play cards close to their chest or never reveal self-interest. A wise person observes with detachment how people manage intentions and interests. They assess and proceed, remaining receptive to unfolding intimacy until manipulative intent becomes repeatedly evident in the other person.

DETECTING INTENTION IN OTHERS

Dishonorable intention is more transparent than commonly believed. Younger, naïve, or less experienced people incorrectly think their intentions are undetectable. Seasoned leaders, people in positions of authority, celebrities, people with a highly developed emotional quotient, and wise people from all walks of life have highly developed intention-detecting skills (also known as a "BS radar").

Manipulative, undisclosed intention works only a few times, even with trusting people. Eventually, the manipulator is exposed as being untrustworthy. Regaining trust becomes difficult or impossible. Impaired relationships, character damage, and lost opportunities are the consequences. The best rule of thumb with strangers is to retain the assumption of good intentions but observe and probe intentions and interests more deeply.

CONSCIOUS INTENTION REINFORCES GROWTH TO POTENTIAL

People are equally adept at sensing honorable and dishonorable intentions. When they sense intentions as honorable and your actions prove it, they trust more willingly, making a relationship possible. You will feel better about yourself and your impact on others. Your quality of life and career will improve. Confidence and motivation for your potential and wisdom increase proportionately to conscious intention. As explained in the other chapters, you will be motivated to attempt practices requiring more effort, such as cultivating consciousness, deepening intimacy, and selecting your psychological state. Sustaining the practice of conscious intention is worthy of celebration.

CASE STUDIES

NAOMI AND LOUISE

Louise is Naomi's mother-in-law of twenty-two years. She is a loving mother to Naomi's husband, Jeff, and a doting grandmother to her three grandchildren. Unfortunately, an edgy emotional void existed between Naomi and Louise for two decades, preventing them from developing a loving relationship. Naomi attempted to ignore it through kind words and by including Louise in family activities, but nothing worked. Her resentment toward Louise grew, even though both women shared love for Jeff and the three kids. Naomi's best friend heard "Louise stories" for years. Frustrated, she finally told Naomi, "You play your cards so close to your chest that it's hard for anyone to know what's happening inside you. Imagine how hard it must be for Louise to trust you. Why not open up about your life, hopes, and dreams? Show her who you are."

Jeff agreed when Naomi asked his opinion. He explained Louise's cautious nature more thoroughly than he ever had before. Naomi began sharing hopes, dreams, and especially her intentions with Louise. Change came gradually instead of in a climactic moment, but their relationship vastly improved over a period of two years. They looked forward to time together and laughed frequently, something rare in the twenty years prior. Naomi told Louise one day that she enjoyed her, something she never dreamed of saying. Louise responded, "Because now we understand each other."

INTENTION IN LEADERSHIP

In my first book of this series, *The Potentialist: Your Future in the New Reality of the Next Thirty Years*, I explained how I became a Potentialist. At the age of twenty, I took a summer college elective course named "Introduction to Personality." A charismatic young instructor introduced our class to the then-revolutionary idea that success is the pursuit of our best selves rather than the acquisition of money, fame, power, social status, or an attractive appearance. He offered a bibliography of human potential in which I eagerly immersed myself and have never stopped exploring. That began my lifelong devotion to human potential, starting with my own. It altered my life's course and shaped who I have become.

Along the way, I discovered the central role of intention in wisdom, the art of living well. My earliest notes, recollections, and journal entries repeatedly observed that doing something for honorable reasons dramatically increases the chances of good results, and doing something for dishonorable reasons increases the chances of poor results. As I understood the law of cause and effect better, it was evident that chance had nothing to do with it. I was promoted to a manager's job at twenty-one and have held leadership roles since. My belief in human potential defined my approach to leadership. I discovered that people enthusiastically support intentions when they are clear, honorable, transparent, and aligned with their interests and those of company stakeholders and society.

Good fortune has led me to found, co-found, or play a significant role in many successful companies. That includes the health plan giant Anthem (now Elevance Health) and the insurance brokerage Acordia (now part of USI). Both companies had improbable beginnings but became New York Stock Exchange–listed industry giants within ten years, thanks to a talented, motivated leadership team and dedicated employees. Management gurus of the time, including author Tom Peters and renowned futurist Dr. Alvin Toffler, found our success fascinating, noting that broad-based teamwork and stakeholder support fueled the remarkable success. I have repeatedly applied the lessons of those ten years and shared them with others.

Leadership and wisdom (the art of living well) share a simple formula: start with an intention that serves you, others, and life equally—and make sure your intention is clear, unwavering, and apparent to everyone.

WISDOM STORY: ALIGNING INTERESTS FOR THE GREATER GOOD

John Mutz had a storied career as a successful, highly respected Indiana businessman and public servant as a legislator, legislative leader, and lieutenant governor. After two long and successful careers, he was asked to take on a third as the CEO of Lilly Endowment, a large and prestigious endowment fund. His business and public service experience prepared him well for a role that could significantly impact the lives of many Indiana citizens.

Endowments are legally required to give away 5 percent of their assets annually. Lilly Endowment is funded by Eli Lilly stock. The stock was doing exceptionally well, and the endowment had considerable funds to invest in good causes. John understood from his legislative experience that government investment is more successful in tandem with private investment. Also, he knew that involving local people increases the odds of success of well-meaning programs. He believed these principles could improve endowment giving.

John asked his staff for their ideas. One staffer's idea especially resonated with John's thinking: the endowment could partner with community-based foundations in Indiana that raise local funds for community development. There were several such foundations in the state, and endowment fund-matching partnerships could spur the creation of many more. John believed this approach could multiply the impact of endowment funds, spurring economic and social services development throughout Indiana.

John demonstrated humility and wisdom by seeking his staff's ideas and insights. He looked beyond the traditional approaches to align interests with local citizens willing to invest in their communities. He remained resolute on leveraging endowment funds through local fundraising matches when it initially met resistance. The program was wildly successful. Today, there are more than 180 Indiana Community Investment Funds, one or more in every Indiana county, with total assets under management exceeding $4.5 billion. It is a remarkable achievement that would not have happened without John Mutz's wisdom to see the more significant opportunity by aligning intentions and interests. The people of Indiana and the Lilly Endowment will be beneficiaries of John's wisdom for generations to come.

CHAPTER 7

Learning States

LEARNING STATES DEFINED

PSYCHOLOGICAL STATES

At every moment, we are in one or more psychological states. We don't think about it unless we feel out of sorts or need to be in a specific state, such as to attend a special event or meet new people who could be important to us. We all know how to change psychological states to adapt to different life situations without thinking about it, or at least not for long. Few people are aware that we can use this psychological state-changing skill to accelerate INME development and progress to our potential and wisdom.

LEARNING STATES ORIGIN

Learning states are specially designed, functional, psychological states intended to promote psychological development. They are based on a technique developed by Carl Jung called "Active Imagination," which he based on much earlier works by Saint Ignatius. An easy way to think about active imagination is "Visualize to Realize." There is solid science behind the power of this technique in performance psychology and neuroscience, as well as practical examples all around you. When effective visualization is utilized, the brain is tricked into thinking that the person actually performed the imaginary exercise.

You have likely heard the homespun wisdom that "We can only be what we imagine." There is truth in this simple idea. The twentieth-century Broadway play and movie *Pygmalion* tells the story of a working-class English girl transformed into a socialite through her imagination and that of her mentor. For years after, the "Pygmalion effect" was a term used to describe a leadership technique based on this idea.

Leadership involves motivating a team of people to imagine what you can accomplish together. Jung took imagination a step further. By consciously and intentionally imagining or visualizing events, situations, or goals with intensity, you can better understand things invisible to your five senses, such as your psychological body, and you can, to a degree, program yourself. Another discovery from Jung's active imagination work was the ability to return to a specific active imagination experience and pick up where you left off. This technique enables a new method of self-directed psychological development. A Jungian teacher and mentor taught me this invaluable skill. In this chapter, you will learn how to use it to shift your psychological state in ways that accelerate your growth and materially improve your quality of life.

DEFINING QUALITIES AND FEEL

Learning states are identified by distinct qualities and intuitive (gut) feel, the ability to choose the state at will, and the growth and insight they offer. Learning states are distinct from emotional states, such as irritation or enthusiasm, and states of mind, such as analytical or pensive, that arise from the Ego. Each learning state has distinct feelings associated with it, but those feelings differ from emotions. Learning states are experienced through the INME that connects you to the Soul function and the innate growth mechanism guiding you to your potential. The experience is pleasurable or even ecstatic. We experience learning states through the gut or diaphragm instead of the brain. Once you grasp the nature of learning states, you will easily identify them and differentiate them from emotional states or states of mind.

Jung defined the active imagination process and goals and took it to a previously undefined level, but the original concept originated from an ancient archetype. Creative artists, scientists, physicians, philosophers, religious scholars, and innovators used similar techniques for centuries to break the limits of daily life tunnel vision. Imagine how crude life today would be without their reflections, contemplations, and inspirations over millennia.

Learning states include Solitude, Awe, Neutrality, Gratitude, Observation, Reflection, Contemplation, and Intimacy. These eight distinct learning states are useful and easily recognizable, although not the only ones available to us. Entering a learning state has similar qualities to meditation. The Ego and Shadow are quieted, and the INME is activated to engage the Soul, but also the Ego and Shadow as needed. Entering a learning state, like meditation, is a conscious choice to become open and learn. Unlike meditation, we can simultaneously conduct most daily activities while in a learning state.

You can transition from one learning state to another. For example, you may be in a state of Solitude and become inspired to Contemplate world matters. You may choose to enter a state of Intimacy when you encounter a friend or loved one. You may be in a state of Awe watching a sunset, then transition to Reflection to address something troubling you. You might be agitated at the beginning of a hike with a friend but transition to a state of Neutrality to offer wise counsel on your hiking buddy's domestic problem.

LEARNING STATE "ADDRESSES"

Eight learning states are described in this chapter. Each state's distinctive qualities (functions) and unique, intuitive (gut) feel are identified. They become the state's "address," like a postal code or internet URL. Memorizing and visualizing each learning state's address imprints it in memory, allowing for a quick return.

This visualization process may sound strange at first, but it is a highly effective way to turbocharge your performance. It is easily learned because you use it unconsciously hundreds of times daily. For example, you practice a sport like golf, tennis, or running by visualizing how to do it right and imprinting how it feels when you do it. When you perform routine tasks like laundry, loading a dishwasher, packing a bag, or driving, each has certain functional qualities and feelings attached. Before long, you can multitask almost all of them because you develop mental and physical muscle memory that frees your consciousness for other things. Similarly, you already use quality/feeling addresses to create a desired psychological state. When you plan a special event like an anniversary dinner or a holiday gathering, you visualize the past "address" of the look and feel you desire and duplicate it.

Reflect on these processes in your life and observe how skilled you already are. Now, read through each learning state. Memorize the feelings and qualities of each. Then, recall and record your experiences with each state to imprint them in your memory. Writing, reading, and rereading states' descriptions and your notes will help you "memorize" their addresses, making it faster and easier to return to each one.

SOLITUDE

"Solitude is the doorway to the infinite."

—SHARAD DESAI

The accelerating pace of change and life scatters our attention into thousands of disconnected fragments, and subtly, unconsciously, we become progressively more fearful, anxious, and exhausted. The Ego and Shadow relentlessly and misguidedly drive us harder and faster to keep up to ensure safety and survival. Their message is to become safer by working harder, being wealthier, having more possessions, being more popular or influential, or physically attractive. They never shut up and always demand more.

We choose the learning state of Solitude to break the Ego/Shadow's fierce grip on us so that we can collect and restore ourselves to reality and meaning. Solitude replaces chaos with calm to see the world and ourselves more clearly. It is an exquisite, silent, distraction-free rapture to observe and connect us to all that is. Solitude is entered by the intention to commune with the Soul's eternal truth that heals and strengthens us. It is fulfilling, loving, and inclusive, devoid of the loneliness that arises from the Ego.

Solitude does not require physical isolation. You can be engaged with others in conversation while simultaneously in peaceful Solitude. If you are in an agitated, competitive, or negative group, Solitude will calm you and the group without a word, allowing you to emanate calm that draws others toward you. Try entering Solitude to experience that effect. Entering Solitude at will is an essential, immediately rewarding evolutionary adaptation to the twenty-first century's increasing pace of life.

Entering Solitude at will while engaged with others may take some practice. It took me several months to develop and hone this skill. For practice, I imagined being at a party in an open field surrounded by trees. Partygoers were laughing, talking, and enjoying themselves. I was also engaged, but then quietly removed myself to a spot inside the forest nearby to become invisible to other partygoers. I was a few feet away, seeing and hearing everything. I paid attention to how that experience would feel. I practiced it many times until I could remember it in seconds, including the feelings and qualities of being present and separate. Then, I began to practice it in groups and one-on-one conversations.

In Solitude, being present and separate does not cause you to disconnect from others, which would prevent intimacy. Instead, you become more engaged and more intimate, not less. This occurs because you are engaged from multiple perspectives, not one. Imagine seeing the Grand Canyon from the river below and from the rim above simultaneously. That's what it's like to enter Solitude while engaging with others.

In public places, when not engaged in conversation, I seek Solitude by imagining I am in an enormous, soundless room with a concave ceiling and no furniture, similar to Rome's Pantheon. I become an integral part of this silent, still place. I wait without expectation, and timeless truth always arrives. One of my most profound insights came to me while I was in a state of Solitude on a crowded flight from London to the United States, actively imagining myself in that vast room.

Many people discover Solitude in public places or unexpected events. A loved one's life-changing Solitude experience occurred on the tourist-packed streets of Pompeii, where she healed a decades-old emotional wound. A friend found peace from heartache in a small child's soothing words on a beach. A mentor discovered his life's work in Solitude in a physician's office after receiving a macular degeneration diagnosis that ended his first lucrative career; in despair, he discovered his life's meaning in a new career.

Solitude often leads to another psychological state to investigate, learn, and resolve matters important to our development. Perhaps Gratitude to forgive without rancor or resentment. Maybe accepting yourself and others through Reflection. Some people discover life's abundance through Contemplation. Follow these urges. Your Soul function guides you to new levels of perspective and wisdom, opening the door to life's meaning. Walk through the door.

Write a vivid description of your Solitude experiences. Describe what you found or hope to find if you have yet to experience Solitude. This will help you to enter Solitude at will in seconds.

Solitude should be easy; hopefully, it is for you. Unfortunately, the Ego and Shadow may interfere, making it challenging, even seemingly impossible. If Solitude proves elusive, you likely have not yet discovered how to call upon your INME to quiet down your Ego and Shadow. Please do not give up; you can find your way with patience and determination.

AWE

"He who can no longer pause to wonder and stand rapt in awe is as good as dead; his eyes are closed."

—ALBERT EINSTEIN

Solitude detaches us from the world to commune with the Soul function one-on-one. Awe disconnects us from the mundane to commune with creation's magnificence. Both are forms of Soul communion, like all learning states. Most people experience Awe but relegate it to an impressive moment instead of an opportunity for perspective or inspiration that leads to their potential and wisdom.

Awe sometimes comes to us when we least expect it rather than being sought. Likely our Soul inside knows what we need that Awe can fulfill. Other times, we seek Awe when troubled, alone, or searching for healing. Feeling resentful or angry can cause us to crave the forgiveness, peace, and beauty accessible in a state of Awe. If you are going through a bad time or are troubled, connecting with nature and its many awe-inspiring sights is a great place to start recovering. Many wise people are moved by the tranquility, peace, and timelessness of a golden sunrise, a sunset, the blue hour's quiet that follows, or an after-dinner stroll. How long has it been since you have done so? Making a practice of beginning or ending your day with nature is a sure way to open yourself to the Soul.

Almost everyone is drawn to Awe because it allows us to shed the mundane and experience a magic moment when we embrace the magnificent. Awe can provide a decisive moment of inspiration and learning, like aha! moments and timeless truths in the Soul's voice. We share timeless Awe experiences with our ancestors: the first sight of the Grand Canyon, a mountain summit, the achievements and mysteries of ancient civilizations, an intricate flower, the miracle of water, the welcome optimism of a bright sunrise, stars on a clear night, a child's laugh, a puppy or kitten's playfulness, a favorite exercise or game, a warm fire, the beauty or gentleness of someone we cherish, or reaching a seemingly impossible milestone.

Pleasure travel is the pursuit of Awe, a blissful feeling of communing with the Soul through a place or event. People worldwide prioritize travel in pursuit of Awe. We labor and save for years to afford Awe-inspiring excursions and vacations to new and exotic places. It is a societal archetype as old as humanity. A worldwide travel and tour guide industry exists to pursue Awe. They could more accurately be named Awe Agents. In a flash, the learning state of **Awe can stimulate transformations through aha! moments and transition to Intimacy with all things, the most profound connection possible between us and creation.**

We can enter a state of Awe at any time or place. Opportunities surround to see, touch, smell, and feel. Take a walk. See the flowers, plants, trees, birds, animals, and the sky beyond. Hold someone you love closely and look into their eyes to see their Soul. This is the gift of Awe, the easiest of all states to enter and enjoy.

NEUTRALITY

The first few lines of what is commonly known as the Serenity Prayer eloquently capture the feeling and function of the learning state of Neutrality:

"God grant me the serenity to accept the things I cannot change, courage to change the things I can, and wisdom to know the difference."

Neutrality lifts us above the egocentric world of for-or-against, on-or-off, and this-or-that to the quantum dimension of neither-and-both. Chapter 3 explained the natural laws of cause and effect and polarity or tension of the opposites. One of their collective functions is safe expansion and advancement. That is, change with built-in quality control. Advancement, expansion, and innovation are offset by adversity, resistance, and unanticipated consequences in order to achieve stability in forward movement.

Neutrality is the state uncommitted to advancement or resistance, one argument or the other, seeking creative resolution. In Neutrality, we accept natural ebb and flow, polarity, and adversity as the universe's timeless order instead of taking them personally or feeling victimized. In Neutrality, we can accept roadblocks to the resolution of problems while remaining confident that someday the issue will be resolved. Think of Neutrality as a counsel of wise, patient, unbiased, fair judges revered for their wisdom. Thirty years ago, a legendary federal appellate court justice eloquently described Neutrality to me as the state he seeks when deliberating.

Neutrality heals like an antacid on an upset stomach, offering peace and contentment. We sleep better. We are calmer. It can restore balance if we tip toward an extreme or become distressed by polarity, extremism, or conflict in contemporary life. Grounded in the INME, we transcend division, judging, or Ego-driven attempts to "fix" everyone and everything to a state of perfection. Seek Neutrality if you become upset or angered by nightly news or pointless division among friends, family, or factions in the country. Enter a state of Neutrality to consider a memory or event with a negative emotional charge like resentment by replaying it as an observer, watching it play out from above. Observe events with a detached perspective like a detective, a judge, or a diagnosing physician. Look for unfolding truth beyond immediate facts. With a neutral perspective, we accept human limitations while motivating ourselves to work relentlessly to improve ourselves and the world around us.

GRATITUDE

We experience the gift of life through Gratitude. Sometimes, we feel and express Gratitude for no apparent reason. Other times, a memory or an external event reminds us to be grateful. We may feel thankful when exiting another learning state, such as Awe, Solitude, or Intimacy.

Gratitude is powerful in its presence or absence. When we are grateful, we're grounded, satisfied, and prepared for what lies ahead. Gratitude has restorative power. We sleep better and are more hopeful and resilient. Each experience of Gratitude brings us closer to our potential and capacity for wisdom.

Happiness and Gratitude are inextricably linked. You cannot have one without the other. Ingratitude originates in the Ego/Shadow's insatiable demands, cognitive disorders, and illusions. Unhappy people live under the cold, wet blanket of Ego resentment, victimization, and inability to connect with the Soul. It is impossible to feel resentful, victimized, sad, angry, or pessimistic in a state of Gratitude. That alone is a reason to enter it daily. Gratitude gives endlessly, asking for nothing and shimmering with potential and joy. Keep a journal of the things for which you are grateful. It's a great pickup for a rough day. Try Gratitude if you feel lost, discouraged, angry, resentful, victimized, or dominated by other negative emotions. If you or someone you know cannot feel Gratitude, seek the support of a therapist, spiritual adviser, or a wise, trusted person.

OBSERVATION

We enter the state of Observation to expand our perspective by seeing life and ourselves as others see us. Have you ever been part of a group but remember it as if from a bird's-eye view? In the Observation state, you are both an actor in the scene and a neutral witness. Many people have the observer experience but file it away as weirdness like déjà vu or an aha! moment. The Observer is often the debut of your INME, the first time you visualize and feel it. You can observe yourself and situations neutrally without judgment or expectations because your INME displaced your Ego and consigned it to its proper role. The saying "I wish I could be a fly on the wall" describes part of the Observation state. With practice, you can be both yourself and the fly.

A beautiful metaphor is the painting *A Sunday Afternoon on the Island of La Grande Jatte*, by the famous impressionist Seurat, which shows people picnicking by a lake. If you look closely, you see nonsensical dots but no coherent image. But take a few steps back, and you see an unforgettable, beautiful scene. Taking a step back from emotionally charged situations in the Observation state provides the same perspective. This example illustrates how using active imagination to become an observer is quite powerful.

One way to initiate the Observation state is to ask yourself, *What is happening here?* and then wait for your intuition to provide an interpretation or a picture. If too much time lapses, memory will distort observation. Observations can be surprisingly straightforward and perceptive. The Observation state exposes what should have been obvious before such as intentions, miscommunications, Shadow functions leaping to attack and defend, and Egos struggling to dominate.

In the Observation state, we may notice ourselves and others behaving wisely, empathetically, and intimately. This event allows us to see the INME in ourselves and others and bring it to the forefront of our experience, accelerating INME development. With practice, the Observation state provides astonishing insight and growth.

REFLECTION

Reflection allows us to perform a wellness check on the psychological body a few, or many, times daily. In Solitude, we commune one-on-one with the Soul. In Awe, we experience the Soul through its creations. In Contemplation, we examine the world outside to understand its reality. In Reflection, we explore the inner world of our psychological body and behavior. Even very young children make behavioral changes to improve relationships or accomplish objectives, suggesting that children instinctively reflect and have a functioning INME before they are consciously aware.

Every responsible adult reflects to some degree, but in the state of Reflection, we have an elevated experience. The stakes to master Reflection are high. Without it, we stagnate and become unhappy, bitter, and alienated. Relationships and social standing deteriorate because we fail to adapt to their requirements. With Reflection, we understand our reactions to specific challenges or people and make necessary changes.

Some people avoid Reflection because they fear facing themselves will bring harsh recrimination and self-loathing that they associate with the Ego and Shadow. That won't happen because Reflection is a state of the INME, which reveals, assesses without judgment, learns, and adapts. **If you struggle with judging yourself, you are trapped in the Ego's cognitive distortion of perfection, which does not exist in nature and cannot exist in you.** Enter a state of Solitude, Awe, Neutrality, or Gratitude to engage the INME and break the Ego's control before attempting Reflection. In a state of Reflection, you learn to be your own neutral, empathetic observer.

Therapists, analysts, and wise mentors serve as neutral, empathetic observers to reflect insights we cannot see ourselves. Effectively, they act as our INME until our own becomes robust enough to engage in Reflection; even a strong, fully formed INME benefits from an independent observer.

"I will do my best, to be my best, and leave the world and the people I meet along the way a little better than I found them" is a grounding maxim for Reflection. It engages the INME and dislodges the Ego's demands. Only your INME knows when you do your best to be your best; let it guide you.

CONTEMPLATION

Contemplation is the opposite of Reflection; it is an outwardly focused, broad inquiry into the functions of life, nature, the universe, and our role. Chapter 2 and later chapters introduced life's unavoidable or existential questions. The learning state of Contemplation is where most of those questions are answered. However, we may need to shift to other states like Neutrality, Solitude, or Reflection beforehand. Contemplation expands the perspectives required for wisdom. We wonder, imagine, ask, and answer open-ended questions. We welcome insights without bias or judgment and accept life's uncertainty.

We may ask fundamental questions about life and our role in it: *How did we get here? Where are we going? Does my life and how I live it matter? What will I experience as life ends? What might life be like in a decade? Why do we struggle to do the right thing for ourselves such as health, wealth, wisdom, success, intimacy, and relationships, etc.?* We may wonder why things are the way they are, or we might wonder why we wonder.

If your answers are consistently negative and dystopian, shift into Reflection and ask: *What compels me to choose or predict adverse outcomes when definitive answers are impossible?* Contemplation opens us to new ways of looking at life and its meaning. We visualize humanity's long journey that brought us to this moment. We imagine where the Soul's aspiration and change forces might take us. We envision how our lives and actions can positively alter the future. We learn to love and treasure the experience of life. Nothing combines Awe and Contemplation better than sitting under the stars, hiking in nature's artwork, or immersing in a pool.

INTIMACY (OR KINSHIP)

We choose the state of Intimacy to become one with our true self INME, Soul, other people, living things, and inanimate objects. The function of Intimacy is connection solely for the joy of being together instead of satisfying other motives or needs. Intimacy is one of nature's most potent forces and life's greatest gifts. It deserves time and attention to explore; the following chapter is devoted to intimacy.

SELECTING AND SWAPPING PSYCHOLOGICAL STATES

FROM UNCONSCIOUS AND HAPHAZARD TO CONSCIOUS AND DELIBERATE

We choose learning states instead of being seized by them like emotional states or moods, and also to escape the Ego's tight grip. Until now, you have likely swapped psychological states without consciously deciding to do so. But sometimes, you make a conscious choice to shift your psychological state. Perhaps you are in an emotional state, feeling irritated or angry for some reason. You attempt to shake the aggravation, but it lingers. Then you remember that you're the presenter at an important meeting scheduled to start in a few minutes. Attending the meeting in an angry state would be disastrous. You quickly refresh your memory on the subject matter and shift your psychological state. By the time the meeting begins, you are amiable and professional, ready to present.

You likely also shift your state in your personal life. Upon arriving home sad or discouraged after a difficult day at work, you realize your spouse or child is upset and needs your empathy and love. You immediately shift to a state of Intimacy to meet your loved one's needs. In both examples, the previous angry or sad state likely does not return, even though its cause remains. This example offers two important lessons. First, you already know how to change states consciously and deliberately. Second, you have a functioning INME because only it, not your Ego or any other function, can swap to the learning state of Intimacy. The skills you need to make these shifts are readily available to you. Now all you need to do is imprint in memory the qualities, feelings, and addresses of each state and intentionally engage your INME to select a state.

COMPUTERS AND SMARTPHONE APPS AS METAPHORS

Early computers executed one simple program at a time. Later, more complex programs that could be operated simultaneously were developed by introducing operating systems to select and swap programs.

Learning states function similarly, swapping from background to foreground to meet specific needs or objectives. The INME acts as the psychological body's operating system. It is the only psychological function that simultaneously accesses the Ego, Shadow, Soul, and Conscience and mediates their opposites. As your INME grows stronger, it more easily quiets the Ego and Shadow. That makes shifting to learning states natural and effortless.

LEARNING FROM UNPLEASANT
PSYCHOLOGICAL STATES

Our first reaction to an unpleasant state is to get away from it. But like everything in the psychological body and nature, unpleasant psychological states serve a purpose within limits. They temper optimism with caution and raise awareness of the unintended and unforeseen consequences of creativity and innovation. Resistance is the winter to creativity's summer. React to a negative/resistive/caution state by respecting its warnings and applying them constructively instead of avoiding or suppressing them. Learn all you can from them.

Then, pay attention to the feeling quality. You dealt with something rather than running or trying to bury it. This is your INME in action. Remember, the INME is the real you, and the changes it creates take root inside you.

LEARNING STATE PRACTICES

The following practices help you easily alternate and utilize learning states, even as you navigate busy days. Take them seriously and stay with them until you master swapping states. You will be astonished by what you learn and achieve. Your Ego will no longer run your life; you will become inseparable from the real you INME.

CALM THE EGO

Ego calming is essential to your growth and quality of life and the first step in optimizing learning states. Many people find quieting their Ego frustrating and exhausting. If that's your experience, don't give up. Once you're more easily able to calm your Ego, selecting and entering the various learning states becomes an effortless habit, like muscle memory in sports and fitness. The following methods work for many people; some are ancient and proven. Additional Ego-calming practices are described in Chapter 4, "The Psychological Body," and Chapter 5, "Consciousness." Find the practice that works best for you and use them as needed.

Method 1: Out with the Bad Air, In with the Good

1. Close your eyes.
2. Take three to six deep, slow breaths.
3. Visualize exhaling dirty, smoky, harmful air and inhaling clean, positive, clear air.
4. Repeat until your mind is calm.

Method 2: Ritual Rinse

1. Go to a room or a place outdoors with a water faucet, like a restroom, kitchen, or garden.
2. Turn on a low-pressure flow of cool or lukewarm water, whatever feels best.
3. Close your eyes and listen to the calming sound of water for at least ten seconds, longer if needed. Be patient and let the sound penetrate deeply.
4. Very slowly rinse your face, hands, and arms while breathing deeply.
5. Imagine a time, maybe as a child, when you felt one with the water as if embraced by it.
6. Repeat until your mind is calm.

Method 3: Recall a Previous Shift of Psychological States

1. Perform Method 1 or 2.
2. Remember exiting a negative mood or a demanding ego's list of things for you to do.
3. What did you do after? Address a loved one's needs? Relax? Read? Nap? Walk?
4. Remember your feelings and the situations before and after the change in your psychological state.
5. Congratulate yourself for running your life instead of allowing your Ego to own it.

Method 4: Solitude

1. Perform Method 1 or 2.
2. Select a calming and collecting activity that you crave, even if you continue to go about your day. Almost everyone has these activities, even if unaware until asked. Examples include:
   - Physical: yoga, stretching, walking, running, biking, swimming, pleasure driving
   - Household: laundry, light cleaning, lawn mowing or trimming
   - Hobbies: crafts, woodwork, knitting
   - Soul-receptive mind releases: journaling, inspirational reading, meditation, sitting quietly
3. Visualize your psychological body, multitasking to engage the INME and calm the Ego.
4. When your mind feels calm, follow the steps below to select, enter, and use a desired psychological state.

IDENTIFY YOUR PRESENT PSYCHOLOGICAL STATE

Ask yourself:

1. What is going on with me right now?
2. If someone were watching me, what would they see?
3. What is my psychological state at this moment? (Remember, you're always in a psychological state.)
4. Did an external or internal event cause my current state or is the cause unknown?
5. Is this state a recurring pattern? If so, why?
6. What do I wish to learn at this moment?
7. Jot down a few notes describing the identified states the first few times. Review when you have six to ten events. Did your descriptions become more accurate? Did patterns emerge? Insights?

CLARIFY YOUR GOAL

Focus on what you want to achieve or learn from a learning state. Is there something you hope to understand or clarify? A new perspective? Relief from negative thought patterns? A greater sense of connection? Choose one goal to begin.

SELECT A DESIRED STATE

Select the learning state you wish to enter using the descriptions, feelings, and qualities described in the eight states summarized in the following table. Match what you hope to achieve or learn (goal) with the state that most closely matches the goal.

State	Function	Focus	Feel/Qualities
Solitude	Connection to the Soul	Inward	Peace, Quiet, Vacuum
Awe	Connection to Nature	Outward	Exhilaration, Wonder, Expansion
Neutrality	Detachment	Upward	Equilibrium, Objectivity, Detachment
Gratitude	Perspective	Upward	Humility, Abundance, Warmth
Observation	Insight	Inward and Outward	Perspective, Patterns, Clarity
Reflection	Insight	Inward	Behavior, Intentions, Clarity
Contemplation	Perspective	Outward	Wonder, Inquiry, Order

MAKE THE SWITCH

1. Focus your attention on your chosen learning state. Feel it as deeply as possible.
2. Enjoy being in the chosen state and accomplish your goal.
3. Simultaneously absorb the unique qualities and feelings of the state.
4. You will exit the state at some point of your own volition or because you are interrupted. You will involuntarily find yourself in another state. In time, involuntary exits will diminish.

Note: If you become distracted, repeat the Ego-calming exercises, and begin again.

REFLECT ON THE EXPERIENCE

1. Was it pleasurable?
2. Was there a sense of accomplishment?
3. Did you achieve your learning goal?
4. Were you ready to exit the state? If not, what interrupted you?
5. You may find it helpful to end a state by shifting to Gratitude for the experience.

SHIFT PSYCHOLOGICAL STATES IN A GROUP

Swapping learning states in group settings begins by asking yourself, *What is happening here? Are we gathered solely for the joy of being together or another purpose?* You might choose to enter Gratitude for being with these specific people depending on the circumstances. Or you might select a state of Reflection: *These people tend to bring out my less desirable qualities rather than my best. Why? Should I be spending time with them?* Or you might enter Observation: *What can I learn from watching us all together?*

CASE STUDY: CONSCIOUS STATE-SHIFTING

My brother and sister-in-law were visiting me in Arizona from their home in Texas when the COVID-19 pandemic struck in 2020. It was an unnerving time. Events were canceled. Businesses closed. Communities were placed on lockdown. Little was known about the virus and how society and the healthcare system would cope. Conditions changed daily. Life as we had known it ended abruptly, with no end in sight. My brother and sister-in-law were increasingly concerned about being away from their children and grandchildren in Texas as conditions deteriorated. Flight schedules were uncertain. Contagion risks in planes and airports were undefined. Long-distance car travel was equally challenging because their two-day drive required stops at gas stations, restaurants, and hotels, most of which were closed.

As we discussed their options as a family, I shifted to the Observation state both to participate and gain perspective. That led to a transition into Gratitude for how we calmly and collaboratively worked through the problematic alternatives and unknowns. Then I recalled how my grandfather died in the 1918 Spanish flu pandemic, leaving behind a widow and two small children, one of whom was my father. Despite the horror of the COVID-19 pandemic, we were fortunate to have modern, widely available health care, unlike the Spanish flu epidemic that struck an unprepared world. I felt grateful for the people working in government, science, healthcare, and pharmaceutical companies trying to cope with the sick and dying and racing to develop a solution.

During those conversations, we experienced profound intimacy, primarily unspoken. We relived our love and friendship over our long lives and the trust and belief in family we shared. We laughed and planned even though it was a difficult situation rife with uncertainty. We also accepted that we might never see each other again, given the pandemic's mortality rate for people our age. We knew that our intimacy would survive even if we did not. A few days later, they drove home nonstop, finding enough gas stations and fast-food options open to complete the long, arduous drive. Because we'd each successfully quieted our Ego and entered into various learning states during the decision-making process, we said goodbye with unconditional love and gratitude instead of anxiety or fear.

WISDOM STORY:
IT'S WHAT'S AHEAD THAT MATTERS

This is a story by Jeff Weiss, a friend of many decades who is wise, thoughtful, and incredibly open and helpful to others. I asked Jeff to offer a story about his toughest decision. Jeff's wisdom demonstrates what happens when the true self/INME is the decision-maker operating from a learning state instead of the Ego that can never admit a mistake.

My professional career began as a clinical and community psychologist teaching family doctors basic psychology and mental health at a UCLA family medicine residency. I sensed academic medicine wasn't for me, so I joined a private practice of family doctors and specialists, hoping for renewed enthusiasm for psychotherapy.

Our practice thrived, and my schedule quickly filled. I hoped my psychotherapy skills would alleviate my career concerns and asked a mentor and supervisor to guide me. But my discontent grew. It became increasingly unbearable to confine my extroverted, entrepreneurial nature to an office full of patients despite involvement in creative distractions such as a new hospice and a pain clinic collaborating with orthopedic surgeons.

I explored alternative career paths. At one workshop, I met an organizational development consultant working in the high-tech industry. We hit it off and joined forces, which matched my interest in organizational psychology. Private practice psychology allowed me to reduce clinical hours to devote to the new business. Unfortunately, our launch was ill-timed as the United States plunged into recession, costing us clients.

We refocused on the growing personal computing industry. A technology CEO interview about success factors led to a conference named "Why High-Tech CEOs Fail," sponsored by Price Waterhouse. We discovered that attending CEOs were more interested in connecting with other CEOs than our consulting. Like all entrepreneurs, we pivoted, offering monthly CEO roundtables where they could coach each other on strategy and operations, and quarterly conferences to promote camaraderie and the Southern California tech community.

Our business flourished, attracting hundreds of CEOs, and putting us on *Inc.* magazine's cover as pioneers in CEO peer groups. The hard personal decision I faced, however, was leaving a successful but unsatisfying psychotherapy career that had cost me so much time and money to create. I decided that the future was what mattered, not past investments. I trusted my instincts, took the less secure path, and charged ahead. I have never regretted the decision. I have since witnessed other healthcare and mental health practitioners make the opposite decision and sadly live to regret it.

CHAPTER 8

Intimacy

"Enlightenment is intimacy with all things."
—DOGEN ZENJI

What if you could experience intimacy so wholly that your relationships became deep and durable, your encounters with others positive, and your kinship with everything in existence undeniable?

A UNIVERSAL NEED

Inside everyone, buried more deeply in some, lies a craving for an ecstatic connection called intimacy that unites us with others and all things in nature. We must offer what we crave to create an intimate connection. That is, intimacy formed and sustained freely, joyfully, and openly. And a relationship in which we are understood, valued, and loved for who we are. Intimacy is far more than a human experience. Given the macrocosm-microcosm theory, intimacy may be creation's neural network, a force or energy connecting all things in existence with each other and their source.

Humanity will achieve an unprecedented elevation in potential and wisdom when intimate relationships are as common as social acquaintances are today. Every psychologically healthy person craves intimacy; our lack of comprehension and inability to create it is the only barrier we face. This chapter will sharpen your understanding of intimacy, so you know precisely what you seek in love, empathy, or relationships. It will explain the preconditions and practices to increase intimacy in your life and outline why it will be more critical than ever in the decades ahead.

Intimacy's benefits far outweigh the minimal effort required. Your relationships will become more grounded, and the love and empathy you experience will be unconditional and sustainable. Even small increases in intimacy will enhance your quality of life and the lives of others. Instead of searching for love, empathy, and relationships in diluted forms, understand and develop intimacy. Allow love, empathy, and relationships to unfold naturally from intimacy's rich soil. Intimacy is unmatched in enriching daily existence and accelerating your potential and capacity for wisdom. Readers often have a favorite chapter in books; I hope this chapter will become one of yours because of the profound, immediate difference intimacy can make in your life.

INTIMACY IS MISUNDERSTOOD AND MISSING IN MANY RELATIONSHIPS

Intimacy, love, empathy, and relationships are commonly misunderstood, especially among young people. For example, love is often conflated with desire and romance, even though both are fleeting, and neither is love. Empathy, when feigned, is frequently self-serving vanity. This confusion leads to superficial connections that many people label as friendships or relationships when, in fact, they lack the depth of connection inherent to either. Our definitions of these terms are so vague and malleable that we add qualifiers: puppy love, romantic love, unconditional love, best friends, casual friends, social friends, soulmates, physical intimacy, etc.

How can we know what we want or guide our youth when we ourselves lack clarity?

INTIMACY'S POLARITIES ADD TO THE CONFUSION

To comprehend intimacy, we must also understand its opposites, like utility. Intimacy is a connection solely for the joy of being together; utility is any other motive or intention. Similarly, intimacy's oneness is opposed by individuality. Intimacy can unite two (or more) people as if they were one while respecting each other as separate people. Resolving these opposites makes relationships durable and meaningful.

INTIMACY DEFINED

"Intimacy transcends the physical. It is a feeling of closeness that isn't about proximity but of belonging. It is a beautiful, emotional space in which two become one."

—STEVE MARABOLI

The *Cambridge English Dictionary* defines intimacy as "the state of having a close personal or romantic relationship with someone."[1] This definition is shallow. Personal and romantic relationships can be whimsical and fleeting; they may or may not involve intimacy.

The *Merriam-Webster Dictionary* defines intimacy as "Familiarity, something of a personal or private nature, a close acquaintance with something."[2] This, too, is misleading. Familiarity is "relaxed friendliness or the quality of being well known based on long or close association." Long-term associations foster, but do not define intimacy. We can experience an intimate connection with someone we encounter briefly and never see again.

Author Jane Austen captured the disconnect between relationship length and intimacy in *Sense and Sensibility*: "It is not time or opportunity that is to determine intimacy; it is disposition alone. Seven years would be insufficient to make some people acquainted with each other, and seven days are more than enough for others."[3]

Intimacy is frequently and incorrectly equated with sexual intimacy. Most intimate relationships are not sexual, including those with our children, family members, and friends. Many sexual relationships lack intimacy, but intimacy can also be beautifully expressed physically through sex as well as platonic affection.

Intimacy's three distinguishing characteristics define it best: power source, unique experience, and psychological function.

Intimacy's power derives from a common creative origin shared by everyone and everything in existence. We crave connection to our creative origin and other beings and objects who share creation with us; nothing else suffices. Kinship, defined as "from one seed," is an insightful synonym for intimacy.

The intimate experience is unique—the simple joy of being together with another act of creation. Combining intimacy's power source and experience expands the definition of intimacy to: "the joy of sharing the common seed of creation."

This definition explains intimacy's psychological function: to encourage and reward oneness and inseparability from a single creative force and all things in existence. The innate drive to reach our potential directly includes intimacy with everything.

CONTRASTS

Differentiating intimacy from love, empathy, and relationship sharpens the meaning of each.

INTIMACY AND SPECTRUMS OF
RELATIONSHIP, LOVE, AND EMPATHY

Intimacy does not differentiate between individuals or vary in degrees of intensity. Many religious and philosophical texts suggest that all things are equal in the eyes of creation. Regardless of your spiritual beliefs or lack thereof, intimacy's psychological function reflects this same idea. Unqualified, undifferentiated, uniform intimacy—the joy of being together—can exist in our love for our children, friends, a spouse or romantic partner, family, pets, and even some inanimate objects.

As oneness and equality of value represent the Soul's dimension, Earthly life represents the opposite. Everything is individualized, from grains of sand to humans. There are hundreds of thousands of plant species, millions of animal species, and billions of individuals in our human species alone, and gender, herds or tribes, cultures, and genetic codes further differentiate them. Yet all living things share the will to survive, grow, compete, and prosper in earthly existence.

Understanding that intimacy is the opposite of individualized needs and utility clarifies the spectrum that intimacy encompasses. Within that spectrum, we experience the richness of intimacy's Earthly derivatives known as love, empathy, and relationship. Examples of the utility of intimacy include mating, procreation, and security.

The tension between utility and the joy of being together requires conscious mediation. As a result, individual capacity for relationships, love, and intimacy varies considerably based on how much each person's intention originates in utility versus intimacy and their effectiveness in mediating these polarities. This tension exists in every relationship, frequently unrecognized and unresolved, leading to relationship friction and failure.

Consider a happily married couple that lives near one spouse's siblings and aging parents. The other spouse receives a lucrative job offer for a dream job two thousand miles away. Taking the job means only visiting family on holidays and vacations and caring for aging parents will fall on local siblings. Should intimacy and empathy for one's partner or the utility value of a dream job prevail? Overvaluing utility can destroy love and relationships, but the demands of intimacy and empathy can diminish individual achievement and rewards.

Utility-dominated relationships are all around us. Many couples seem to have little to discuss and experience no joy in being together. Others sacrifice intimacy to win power struggles over differences once valued as being complementary. Some people claim to be lovers or friends but behave more like competitors. Some co-workers avoid personal connections throughout their careers. Too often, social "friendships" are full of banter and competition but devoid of substance. This overabundance of utility-driven connections results in a society marked by distrust and suspicion, unsatisfying short-lived love affairs, and superficial friendships and social media connections that fail to address pervasive loneliness.

Fortunately, we also witness love, empathy, and relationships based on the joy of being together. Enduring friendships, marriages, families, and partnerships result from unconditional love and commitment. While these relationships may initially have involved some measure of utility, that purpose no longer exists or matters. Each party is a fully independent, whole individual who chooses the relationship for the joy of being together. These bonds are often described as "close-knit families," "great marriages," or the "best of friends"; the intimacy within these connections inspires us all.

Utility-dominated relationships are more common than intimacy-dominated relationships, in my experience. What are your observations?

INTIMACY AND UNCONDITIONAL LOVE

Without intimacy, love is conditional and never totally fulfilling. Intimacy is the portal to unconditional love; both endure in memory and experience even if you never see the person again. Intimacy and unconditional love survive time and distance, especially with modern communication. The letters between US president John Adams (1735–1826) and his wife, Abigail Smith Adams (1744–1818), demonstrate how intimacy can be maintained and deepened through written correspondence over decades despite long periods of physical separation. Spanning almost forty years, their letters, which began during their courtship in 1762 and continued until John's political career ended in 1801,[4] illustrate intimacy's depth and durability. I recommend reading at least a selection of their best letters.

INTIMACY AND POSSESSION

If you unconditionally love someone, would you want the best for them even if it meant you would no longer have a relationship? Could you be happy for a spouse, partner, or friend who left you if it was best for them? Could you unconditionally love a child who decided that they no longer wanted a relationship with you? Suppose your spouse decides to pursue a career in another state. Could you support their decision, even if it led to divorce or living apart? Intimacy, the joy of being together, has no conditions. Unconditional love means "I want the best for you even if it isn't me." Your love, empathy, and relationships remain conditional if you cannot go so far.

Unconditional love is complicated or impossible for many people, especially younger people who mistake possession for love. Unconditional love is not ownership; instead, it values a loved one's freedom over possession. "I want the best for you, even if not me," repudiates possession and exemplifies intimacy. Possession, originating from the Ego, makes love conditional: "I love you only if you are mine." There is no love in jealousy. Have you ever had a passionate romantic relationship that suddenly became bitter, vindictive, dangerous, or violent? These situations are madness arising from Ego-driven possession that is devoid of intimacy.

INTIMACY AND CONSUMPTION

We often witness jealousy emerge if a spouse, child, or friend grows close to someone else. The offended person feels rejected because they believe love is a fixed quantity. This Ego-driven thinking suggests that love given to another reduces what is available to them. Intimacy, unconditional love, and empathy are not subject to the law of consumption but to infinity, from which they are derived. The more we give, the more we have available to offer.

INTIMACY AND CONDITIONAL LOVE

Conditional love is intimacy that has been diluted or eclipsed by utility. I love you because . . . you are my boyfriend, girlfriend, child, friend, physically desirable, wealthy, intelligent, powerful, etc. There is no "because" in intimacy. Relationships based chiefly or solely on utility gamble that intimacy will grow with familiarity; more often, contempt grows instead.

Conditional Love	Intimacy
Founded on utility	The joy of being together
Arises from the Ego	Arises from the Soul
Wide spectrum of variations	Uniform
Often stops growing	Grows over a lifetime
Rarely lasts a lifetime	Usually infinite
Can be one-way	Always mutual
Drawn to the outside person	Drawn to the light inside
Threatened by another's love	Celebrates all love
Possessiveness is common	Cannot coexist with possession
Rushes forward on a plan	Unfolds naturally like a flower
Available to all	Available only to the committed

INTIMACY, ROMANTIC LOVE, AND SEXUALITY

Intimacy is often confused with romantic love, which is an attraction for a reason, such as a prospective mate or partner, sexual adventure, fantasy, or some combination of these. Romantic love projects a fiction or myth onto the desired partner instead of discovering and wholeheartedly embracing their reality. Once the fiction fades, intimacy can develop with the real person, but it often does not.

Romantic love usually anticipates or includes sexuality. Mass media glorifies sexuality as an instinctual or utilitarian act, but for most people, sexuality without intimacy produces a temporary physical release and Ego boost but results in disappointment and alienation. Engaging in experimentation and making mistakes are typical, but mindless promiscuity can scar you and make essential self-acceptance more difficult.

Sometimes, sexuality without intimacy occurs but leads to intimacy that develops incrementally over time. This transition may be more feasible as people mature and learn to value intimacy over sexuality.

At their best, romantic love and sexual intimacy unfold naturally from other forms of intimacy, such as friendship and shared experiences and interests. Physical intimacy can deepen trust, comfort with vulnerability, and love without possession. Time and circumstance disappear during physical intimacy; creativity and spirituality become seamless with the physical union, engaging all our senses in a way closest to a state of Awe.

INTIMACY'S ALLURE

At some point in life, most people mistake a feeling of intimacy or sexual attraction for love. Have you ever had an intense, unexpected, unexplainable attraction, knowing the person couldn't be a love interest? Have you ever felt in love with someone and become physically intimate, only to discover that it was friendship you wanted?

Intimacy signals a meaningful connection worth pursuing, not someone's role in our lives. When you experience intimacy, allow the relationship to unfold naturally. **Enjoy intimacy, whatever its form, without forcing any direction, timetable, or socially acceptable label onto it.** The relationship will evolve organically to what it should be. This is an important life lesson to avoid unnecessary heartbreak, disappointment, and failed relationships and marriages.

INTIMACY WITH ANIMALS AND INANIMATE OBJECTS

Many people feel deeply connected to animals, especially pets, and believe their intimacy is returned. Inanimate objects such as an heirloom, a work of art, a favorite sweater, or a coffee cup may touch us deeply. Many people name their cars and houses, and the objects seem to assume a personality fitting the name. Even a sunset or soft breeze may engender strong feelings of connection with the beauty and mystery of life. Anthropomorphism projects human qualities onto animals and inanimate objects. It can be fun as long as we acknowledge and accept their reality.

BARRIERS TO INTIMACY

Certain attitudes and behaviors inhibit intimacy, love, empathy, and sustainable relationships. The Ego interferes with intimacy and unconditional love, injecting utility, possession, and other agendas and criteria. Seek help from a therapist, counselor, friend, or mentor with solid relationship skills if you struggle with these barriers and cannot overcome them alone.

EGOS

We are often drawn to people with strong egos because they tend to be committed, high achievers. Intimacy cannot exist between two Egos or one Soul and one Ego. An Ego can never love you, and you cannot love an Ego. Intimacy requires trust and openness. The Ego attempts to supplant intimacy with utility and often with control. If you are involved with someone dominated by their Ego, they will likely exploit your trust and transparency instead of honoring it. Intimacy requires submission of the Ego to the Soul, and the Ego always resists. Lasting intimate relationships require two people equally capable of intimacy.

FEARFULNESS AND RELUCTANCE TO TRUST

Intimacy is two or more Souls wishing to merge. Most people are capable of intimacy but struggle to overcome the barriers created by fear arising from the Ego's survival function. When you have an opportunity for intimacy, expect your Ego/Shadow to obstruct you with an endless list of fears, such as vulnerability, rejection, unworthiness, and humiliation. In that moment, you must accept those risks or miss a chance at intimacy, one of life's greatest gifts. I find inspiration in Julia Roberts's line in the classic movie *Steel Magnolias*: "I would rather have thirty minutes of wonderful than a lifetime of nothing special."[5]

A few people are so fearful that they become incapable of intimacy, living disconnected from people and nature while surrounded by both. Author Melissa Banks in her book *The Girl's Guide to Hunting and Fishing* describes this human tragedy as "Dante's definition of hell: proximity without intimacy."[6] People whose fear overrides their capacity for intimacy deserve our empathy and kindness. Today, many people in midlife or older who live alone turn to pets because they offer lower rejection risks, fewer intimacy requirements, and more stable relationships.

Even when intimacy seems out of reach, change can happen in a moment. A lonely person might unexpectedly encounter someone around the corner seeking a relationship. A long-standing neighbor might suddenly want to connect. Hope lies deep within us. Intimacy reduces fearfulness and increases trust when it is given a chance.

EXPECTATION OF RECIPROCITY

Expecting others to return the intimacy we offer is self-defeating. Some people will be incapable of loving you as you love them or being as open to intimacy as you are. Others will return all the love and intimacy you can dish out. Accept what develops without expectation or regret and allow what you find to be your guide moving forward. Remember, the most durable relationships form between two people equally capable of intimacy. Sometimes, that unfolding occurs with time and patience. Only you can assess a relationship's potential.

LIVING FOR ONESELF AND LIVING FOR OTHERS

Nature's distribution of personality traits creates a bell curve of those who live for others (serving the whole) and those who live for themselves (serving the individual). Both polarities are unhealthy in the extreme. People's capacity for intimacy parallels this bell curve. Intimacy is impeded by being too strongly predisposed to live for others or oneself. Understanding the polarities and constructively using their tension is essential to healthy relationships.

THE INTIMATE EXPERIENCE

"You know the deep release from life's struggles in intimacy's safe harbor and the painful void when it's not there."
—DEBBIE SMITH

Most of us have experienced intimacy, if only fleetingly. Like aha! moments, you may need help understanding how it happened and how to recreate the experience. You may have confused sexual attraction or romantic love with intimacy. You may have thought of intimacy as only between two people instead of an experience you can have with yourself, a group, an animal, an object, or a profound moment. You may have considered intimacy as one of life's unexplainable anomalies, like déjà vu, instead of a pinnacle of human achievement worthy of deliberate pursuit. You may have dismissed intimacy as a random good feeling instead of connecting to creation, the most powerful force you will ever touch. **Intimacy is fully comprehended only when experienced. Its rapture leaves you without doubts or reservations and wanting more.** Someday, we may have instruments to identify and measure intimacy. For now, we must be content with the experience.

A DISTINCT AND UNIQUE CHANGE

"A connection has emerged between us based solely on the joy of being together, unlike anything I have experienced in its tender intensity. We write, talk, touch, and love in an exclusive space gifted to us for reasons we can't fathom, by means we could never have guessed, for a time and direction we cannot anticipate. I find it exhilarating to find myself thinking about the next moment with you as the last moment ends."

—F. SCOTT FITZGERALD

Please take time to slowly read the F. Scott Fitzgerald quote above again. It is one of the most exquisite definitions of intimacy I've found. While his context was a romantic relationship, the words apply to any intimate connection—a friend, relative, or even a pet if you exclude the writing part.

Intimacy is always accompanied by a distinct and unique change in feeling; something profound is occurring, and you know it. Energy increases alongside an unusual calm, two things you would not expect to experience simultaneously. Hundreds of people have told me over the years that they felt calmer after I spoke with them individually or in groups. That was intimacy.

I did nothing intentionally to calm anyone.

One gentleman described his intimate experiences as gravitas—knowing something important was underway. Other people feel exhilaration alongside warmth and serenity. Still, others say it felt like a third entity entered the room, an aura of love or something less identifiable, like an energy increase, electric moment, intensity, feeling spellbound, or something magical. Some people experience changes inside their bodies arising from their solar plexus or stomach instead of their heads. You may experience intimacy in any or all these ways or a unique way, but there will always be a distinct change in feeling with a beginning and end.

Are these experiences real or illusory, caused by strong emotions from connecting deeply with others? **Intimacy may be the most powerful force we experience in life.** It isn't surprising that it causes unusual physical, emotional, and spiritual experiences. Intimacy enters the room and alters the bodies and minds of the people involved. Intimacy transcends love, empathy, and relationships as we know them, like turning up the volume on a symphony until the music permeates your Soul.

SERENITY AND PEACE

Intimacy may offer a glimpse into the possibility of existence beyond death. A person I greatly admire shared an experience in which he died during a medical emergency. He was retrieved from certain death by the heroic actions of emergency medical professionals. He described his experience of being clinically dead, yet aware, immediately after being resuscitated. His description surprised others by the accuracy and clarity of his recall. He reported feeling only joy and calm throughout the experience, with no desire to return, even as doctors fought to save his life. His description approximates how people experiencing intimate moments never want them to end because they are an end unto themselves, an encapsulating rapture.

A few months later, I met a woman whose PhD dissertation was on life-after-death experiences. I learned that my friend's story is not unusual.

Most of us are unlikely to have such a profound experience. Still, we can pay attention to intimate moments and wonder if unconditional love, peace, and a sense of belonging might define existence after physical death.

INTIMACY WITH ONESELF

We often pursue love and relationships with others, overlooking the prerequisite of intimacy with our own Soul and INME. As you know, accept, trust, and become intimate with your imperfect self, you are better able to do the same with beautifully imperfect others, life, and the world.

I began to understand the importance of intimacy in my thirties. I knew how to be a good friend to others but had not yet learned to be my own friend. A mentor asked me why I drove myself relentlessly at seventeen years old to excel in school, sports, and multiple jobs until I almost died from a bleeding ulcer. Could I understand and forgive that kid? Did I trust that my relentless, irresponsible drive had matured into something I could manage? As I reconnected with my Soul and INME, I learned and accepted myself while continuing to become my best. I realized that opportunities for intimacy with myself and others were endless.

Intimacy begins with us. Visualize yourself as a friend needing intimacy and unconditional love, and then provide them to yourself wholly and willingly. Intimacy with yourself defines the limits of your ability to be intimate with others.

THE INTIMATE COUPLE

An intimate couple expands their consciousness and accelerates their potential by combining the power of aligned souls. Everything is secondary to the joy of being together. They trust each other and themselves, confident in their capacity to overcome hurdles and grow together. They are eager to share self-knowledge and understand how others see them.

The couple rides the waves of change rowing together. They become ever more committed to shared and differing values that draw them to a stronger center. Growth is supported without fear of being left behind or the need to slow down so the other can catch up. Theirs is a union of Souls with little or no separation. They are steadfast, unthreatened, and brave.

They laugh, cry, and talk about everything. Conversation flows effortlessly from growth and evolving awareness. Silence is comfortable and accepted as another way to connect. They speak in shorthand, so a sigh or grunt can become an expository essay. They are in a sacred space without rules and conventions other than their own. Time and distance are irrelevant, and their intimacy grows even if their bodies are separated.

No one wants to break the spell. When life intervenes . . . the phone rings or an unavoidable prior commitment needs attention . . . the intimate moment evaporates, though not its imprint. They will reconnect in an intimate place, at a higher plane, never to return to the starting point.

THE INTIMATE CROWD

An intimate crowd may sound like an oxymoron, but it is the collective Soul at work. The combined power of Souls, a congregation, is essential to every major religion and most spiritual and philosophical schools. Group prayer, singing, and fellowship have advanced civilization for centuries without sacrificing individuality. Intimate crowds make the same commitments as intimate couples but are more challenged to fulfill them. More people, Souls, individual differences, and spectrums of love, empathy, and relationships are involved. Crowds may begin with good intentions but be distorted by one or more members' Ego-driven desire to dominate or pervert the communion of Souls.

Teamwork, whether in business, the arts, sports, entertainment, or other fields, is a form of crowd intimacy that can be remarkably creative and fulfilling. In the twenty-first century, work will become increasingly collaborative, using technologies that enhance creativity, communication, and speed. Learning to experience intimacy within these contexts will become an important skill, especially for young people.

INTIMACY PRACTICES

You are ready to put intimacy into practice and make it habitual. The following steps need not be practiced serially. Once they become natural, the whole process becomes effortless.

SHOW UP!

We have all ruined intimacy opportunities by prioritizing something less important. We've also shown up physically with our minds or hearts elsewhere. And we've all been on the receiving end of people who did the same to us.

I once failed a beloved grandson, losing a precious opportunity for an hour that can never be replaced because I was emotionally disconnected. He recognized what was happening at six years old, and I knew he knew. I swore that day I would never again miss an opportunity when intimacy beckons, and I haven't. Intimacy with another person begins when both show up physically, intellectually, psychologically, and emotionally. Don't miss it.

INTIMACY OPPORTUNITIES

Opportunities for intimacy are everywhere. Like our hunter-gatherer ancestors who walked on rich earth for many generations before discovering that tilling it could end their hunger, we unknowingly walk past an abundance of love, empathy, and relationships that could end our emotional and spiritual hunger.

Every interaction has intimacy potential, including those with people we have never met and are unlikely to meet again. At almost any moment, intimacy can occur or deepen with spouses, partners, family, friends, neighbors, and people we encounter in our careers and social lives. We can experience more intimacy daily by openness to these opportunities. It isn't as tricky as it sounds. You catch yourself the moment before you meet someone or a loved one who walks through the door or gets into your car. You consciously tell your Ego that the priority is your loved one, and your Ego is not to interfere.

If you need help with focusing your awareness and attention on others, return to the collecting and Ego-calming exercises in the prior chapter on learning states.

PAY ATTENTION TO ATTRACTION

We do not question being drawn to someone when there is a utility like appearance, money, fame, power, or social status. Why do we hesitate when we are drawn by intimacy, like a sixth sense? Less predictable attractions may hold the intimacy prize. We can sense fascination or mystique with another person even when the reasons are unexplainable or unjustified. Follow attraction without expectation and see where it goes.

MAKE THE OFFER

Intimacy cannot be forced or controlled, but offering intimacy is almost always healthy, whether or not it is accepted. An offer can be as subtle as "I enjoy our time together" or more explicit as "I would like to deepen our friendship." You will not connect every time, but the attempt alone influences. So be brave. Doing so can have an outsized effect for little effort. Every time we offer intimacy and accept it from others, a positive ripple effect occurs, expanding intimacy within our circles and beyond into theirs. Every person needs and craves intimacy as much as food and water. **Think of intimacy as a healthy contagion spreading outward and altering every person and thing you touch.**

IT'S YOUR DECISION

Increasing intimacy in your life and relationships is an act of free will. Your Soul will urge or suggest an intimate connection, but your Ego and Shadow will interfere, warning of inadequacy, rejection, or humiliation. If you recognize and patiently hold the tension between the Soul/Conscience and Ego/Shadow, your INME will arise to mediate. Your INME will let the Ego/Shadow know that the risk is understood and accepted. This fulfills the essential function of the Ego/Shadow without dictating your decision.

Do you remember retreating from previous intimacy opportunities? Do you regret the decision or feel it was best at the time? Can you identify when and why your Ego/Shadow resists intimacy?

TRANSFORM IMPERSONAL INTERACTIONS INTO INTIMATE MOMENTS

Try this ten-second intimacy practice. Take a breath before you engage someone in what would otherwise be an impersonal transaction. Close your eyes, if only for an extra blink. Shift your psychological state to the learning state of Intimacy. All of this can occur in less than a second.

Then, bring a short phrase into your conscious mind. Some of my favorites include: "See this person, not the role they play," "Let the light in me shine through," "This person is a brother or sister," or "Treat this person like family." The phrase connects you to the INME's intimate learning state and away from the Ego's busy chatter. Double-check your intention, ensuring it is solely the joy of connecting with the other person, not to make yourself feel good. Connect to leave a positive mark on the person's day, even if you never met them before and are unlikely to encounter them again.

Look them in the eyes. Smile. Make sure they know you see them. Engage with them. Ask their name and give them yours. Ask how their day is going. Demonstrate genuine interest by asking how long they've worked in the job, what they enjoy and don't like, etc. Yes, it's small talk, but many people will feel genuine intimacy if you offer it. We all want to be seen and known, even if we are unaware of or deny it.

Please make a special effort with people in jobs typically unrecognized, overlooked, or underappreciated, such as service workers, police, firefighters, healthcare workers, retail, etc. Some will not accept your sincerity and intimacy, but many will, and it is always a positive act. Take two to three seconds after the exchange to reflect on the person and the encounter. Express your gratitude for meeting them and wish them well openly or silently.

INTERPRET THE MEANING OF INTIMATE MOMENTS

Interpreting an intimate experience afterward is as important as deciding to embrace intimacy. Taking that small slice of time to comprehend what you experienced determines how likely you are to pursue intimacy in the future.

This self-inquiry can include questions such as:

- Has the intimacy remained, intensified, or faded?
- Did you grow from the experience?
- Has distrust or doubt arisen?
- Do you have any regrets?

Doubt and regret are natural, so don't let them concern you. They often arise from your Ego/Shadow's attempts to reinstate self-defense and safety precautions that it believes will protect you from the vulnerability required for intimacy. This safety is fiction, of course, since it leads to isolation and loneliness that cause more pain than failed attempts at intimacy. Doubt also results from misjudging an intimate event. Souls connecting can have a Svengali effect, causing a misunderstanding of the event or the persons involved. This can occur when, for instance, the intensity and magical qualities inherent in intimacy are misjudged as falling in love or planting the seeds for a friendship that never develops.

As you reflect on intimacy you've experienced, evaluate it with a broader perspective and increased patience. **Intimate events act like beacons, highlighting life's meaning available to you.** Avoid interjecting a purpose or outcome you desire; remember, those derive from the Ego, blocking intimacy rather than fostering it. Instead, allow the intimate event's purpose and direction to unfold on its own. If the intimate event was a one-time occurrence, accept the gift. If it recurs, find joy in being together without a goal, purpose, or timeline. Make a conscious mutual decision before expanding intimacy to physical intimacy. Unconscious choices can cause injury and heartache. Conscious decisions between two mature adults can deepen and broaden intimacy in healthy ways.

INTIMACY IN THE TWENTY-FIRST CENTURY

Studies consistently show that close relationships are required for happiness and extended longevity. Today, most people tap only a tiny drop of intimacy's ability to improve their quality of life. As life expectancy increases, so will intimacy and its gifts of unconditional love, long-lasting relationships, and empathy.

According to Dr. James Flynn, the average IQ has increased by sixty points in one hundred and ten years, virtually all in abstract reasoning.[7] Imagine the IQ increase possible in the next quarter-century and beyond as medical innovations abound, and our brains link to unlimited computing power and memory. Comparable growth will be necessary and more available to us in human communication and relationships.

People worldwide will have a historic opportunity to live longer and better, perhaps significantly so, due to advances in medicine, science, and safety. More of the world's population will live longer in their peak years of maturity and wisdom. Developed countries and, ultimately, the world will become less crowded as the population declines. We may value each other more as every person is needed to sustain economic growth, and innovations shrink the world and blur our differences.

Worldwide high-speed internet and cellular systems will topple barriers to communication, including language differences, education, financial opportunity, relationship building, and even wisdom. More people will be able to choose where they live instead of their work forcing them into cities or long suburban commutes. Fewer people will need to immigrate as the cloud makes economic opportunity at home possible. Interaction with the internet and other automation will be by voice and thought without any knowledge of computers or a keyboard and screen. Experience-sharing technology will create a near-lifelike substitute for in-person gatherings to collaborate and learn together.

However, as communication advances proliferate, we will need to be discerning about how we use them. Today, more of us are connected than ever; having several hundred or several thousand online connections and followers is increasingly commonplace. Yet, loneliness and alienation remain a significant threat to the mental and physical health of younger and older people alike. Rather than emphasizing superficial connections for their utility, we will need technologies that encourage and support intimacy or we will squander our potential and quality of life.

As we grow wiser, we seek more meaning in our lives. We must experience a sense of belonging in our homes, communities, and world while helping others feel that they, too, belong. **Intimacy is the North Star of the art of living well, our most basic need, our highest calling, and the surest path to potential and wisdom.** Remember that North Star if you feel down, frustrated, sad, or isolated. You will never be alone if you have even one intimate relationship,

Ultimately, you must decide if intimacy will be the priority in your relationships; only then can you move them from troubled and distant to intimate. What if most or all of your relationships expanded to unconditional love? What if you engaged everyone you meet, even briefly, in an intimate moment? It is not your job or mine to change the world, but it is our responsibility to maximize our potential and improve the lives of those we touch. What could the world become if intimacy was humanity's natural state? It starts with each of us.

WISDOM STORY: WHAT'S BEST FOR YOU IS BEST FOR ME

Deems Dickinson was a senior executive with Russ Lyon Sotheby's International Realty in Scottsdale, Arizona, for thirty-nine years, the past fourteen as president and designated broker before his retirement. He is universally admired for his wisdom, leadership, character, and business savvy. Deems tells this heartwarming story:

> My father, Don Dickinson, was my business mentor throughout my long career. After graduating from Arizona State University, my dad and I worked alongside each other for four years as sales associates at Russ Lyon Realty Company. Then, we decided to launch our own partnership, The Dickinson Company, serving the Northern Scottsdale and Phoenix area, primarily selling lots and land.
>
> The opportunity to work together was an exciting time for us both. I learned so much from Dad, witnessing his business acumen and wisdom up close, awed by his experience, knowledge, and judgment. Having his son follow in his footsteps was a dream come true for Dad.
>
> But along the way, I discovered my true calling to be a leader. I excelled at selecting and developing talent and helping them succeed, even though I treasured working alongside Dad in real estate sales for six years. The legendary Arizona real estate entrepreneur Dennis Lyon offered me an opportunity to manage

a newly opened office with his firm. It was an opportunity to move into a leadership role, but it meant dissolving the partnership with Dad.

I struggled with the decision, even though I knew it was the right move. Typical of Dad's wisdom, he also recognized it was right for me. My strengths and desires were leadership and management, and the long-term opportunity with Russ Lyon was bright. Dad encouraged me to accept the position to the detriment of his plans and dreams.

With Dad's encouragement, that decision set the course of my success for the next thirty-nine years. His wisdom to see all sides of an issue, even at personal cost, was an unforgettable lesson. Our lives were richer for it, and he continued to be my mentor throughout my career and life.

PART III

Approaching Potential

CHAPTER 9

Reframing Fear

> "Each of us must confront our own fears, must come face to face with them. How we handle our fears will determine where we go with the rest of our lives. To experience adventure or to be limited by the fear of it."
>
> —JUDY BLUME

WHAT IF RELATING TO FEAR, ANXIETY, AND INSECURITY differently could be a source of calm, confidence, energy, strength, and wisdom?

THE AVOIDED PSYCHOLOGICAL FUNCTION

Wise people I have known emanate quiet, optimistic confidence and courage. They experience fear but do not let it rule them or dominate their decision-making. Fear that is understood and kept in perspective should not be denied or ignored; it keeps you safe. However, unconscious, avoided, denied, or misguided fears allow your Ego and Shadow to dominate your life, making you miserable. Unsurprisingly, most people avoid delving into their fears, anxieties, and insecurities. You can become calmly courageous instead of fearful and anxious by consciously addressing your fears.

Almost everyone occasionally struggles with nondisabling fear, anxiety, insecurity, or vulnerability. This chapter offers a layman's model to understand the fear function, how to avoid being manipulated through fear, and how to transform fear into calm courage. Applying it can improve your quality of life, accelerate your potential, and expand your capacity for wisdom.

This chapter is not intended for those under the age of twenty-one, individuals who suffer from chronic fear, anxiety, insecurity, and vulnerability that interfere with normal functioning and quality of life, or anyone under the care of medical and mental health professionals for these conditions.

Fear is an essential, complex, evolving psychological and physical survival function. Animals have well-developed fear functions, but only humans have the capacity for consciousness, making our fear experience more subtle and deceptive. While consciousness can distort the role that fear is intended to play, it can also turn fearfulness into calm courage.

Most people have been taught little or nothing about fear and how to work with it, so any knowledge gained can be liberating. We have been taught to ignore, deny, or avoid fear instead of understanding it, and the collective Ego in society reinforces this outdated approach. Mass media, political parties, advocacy groups, and marketers specialize in using fear to sell and manipulate.

Many books and a wealth of research on fear and related conditions are available from the perspectives of science, medicine, psychology, sociology, theology, and philosophy. You do not need to read a mountain of materials to have a productive, healthy working relationship with your fear function. The model presented in the following pages lays a foundation for you to get started. It may also inspire you to research other techniques or consult a therapist so that fear doesn't own you or lead you to make unwise decisions.

URGENCY TO CHANGE YOUR RELATIONSHIP WITH FEAR

"This great Nation will endure as it has endured, will revive, and will prosper. So, first of all, let me assert my firm belief that the only thing we have to fear is fear itself—nameless, unreasoning, unjustified terror that paralyzes needed efforts to convert retreat into advance."

—PRESIDENT FRANKLIN D. ROOSEVELT[1]

Fear is humanity's most significant challenge in the next three decades. Unchecked fear has a cancerous impact on you, your loved ones, community, nation, and the world. It spreads quickly and easily, distorting perception and constraining your potential and wisdom.

People seem more fearful today than in decades past. You can sense fear in conversations with friends and family, in social gatherings, in heated political rhetoric, catastrophizing news reports, places of worship, and the tension and uncivil behavior in crowds and on highways. Yet most people live safer, longer, and better than ever. Fear is rising when it should be receding. In Chapter 2, we explored an obsession with extinction as a reason for widespread fear. Now we'll look at fear's function and why people are increasingly susceptible to it.

1. Fear is becoming a dominant theme in contemporary life. Observe its outsized effect on conversations and thinking compared to a decade ago.
2. Fear makes us indecisive because it blocks wisdom. Inaction becomes habitual, leading to missed opportunities. A frightened population weakens an otherwise healthy economy and society.
3. When fear is present, intimacy is absent. Loss of intimacy leads to alienation and the inability to trust required for durable, meaningful relationships.

4. Widespread fear and anxiety manifest in ways that we do not typically associate with fear, such as extreme advocacy, conspiracy theories, and unquenchable desires for security and certainty through fame, money, appearance, social status, and power.

5. Fear can be more dangerous than the underlying threats causing it. Many people today border on panic that could spark mass hysteria, overwhelming individual and societal norms of conduct.

6. In a democratizing society, you have increased power and influence to confront fear. Never in history have individuals had such a megaphone. As goes the microcosm, so goes the macrocosm. Communities and countries mirror the fears of their citizens. Overcoming your fears to become calmly courageous will influence friends, family, community, nation, and the world. So speak, act, and influence.

DEFINITIONS

The terms *fear*, *anxiety*, and *insecurity* are often misunderstood, even though everyone experiences and discusses them.

FEAR DEFINED

The *Oxford English Dictionary* defines "fear" as "an unpleasant emotion caused by the belief that someone or something is dangerous, likely to cause pain or a threat."[2] Dr. Paul Ekman defines "fear" as "the universal trigger for fear is the threat of harm, real or imagined. This threat can be to our physical, emotional, or psychological well-being. While there are certain things that trigger fear in most of us, we can learn to become afraid of nearly anything."[3]

ANXIETY DEFINED

The *Cambridge Dictionary* offers the following definition:

> Generally, [anxiety is] a feeling of worry, nervousness, or
> unease, typically about an imminent event or something with
> an uncertain outcome. In psychiatry, anxiety is a mental con-
> dition characterized by excessive apprehensiveness about real
> or perceived threats, typically leading to avoidance behaviors
> and often to physical symptoms such as increased heart rate
> and muscle tension.[4]

A helpful way to view anxiety is as a collection of undifferentiated,
unresolved conscious and unconscious fears resulting in a persistent
state of unease. To address anxiety, the fears must be isolated and
dealt with individually while simultaneously building self-confidence
to endure.

INSECURITY DEFINED

The *Oxford English Dictionary* defines "insecurity" as "uncertainty or
anxiety about oneself, a lack of confidence."[5] WebMD's definition is:
"Insecurity is a feeling of inadequacy (not being good enough) and uncertainty. It produces anxiety about your goals, relationships, and ability to
handle certain situations."[6]

DISSECTING FEAR

"Thinking will not overcome fear, but action will."

—W. CLEMENT STONE

Most of us were required to dissect animal cadavers in high school or university biology classes. Our curiosity and pursuit of a good grade compelled us to overcome the scariness, smell, and repulsion of cutting into a dead animal. After a few dissections, our apprehensions faded, and the exercise became a mundane routine. Dissecting fears is similar. We must overcome a lifetime of avoiding, denying, suppressing, or cowering, which increases fear's power over us and limits our potential and wisdom. **Fears thrive in darkness, growing into amorphous, undefined monsters. Deconstructing fears brings them into the light, where we can examine them objectively to understand their source, characteristics, and ways to respond.** We come away empowered, ready, and able to face and digest fears in healthy ways, accepting the lessons they offer about ourselves, others, and the world.

The following dissection model uses simple steps, graphics, and metaphors to transform invisible, shapeshifting fears into physical objects that can be studied objectively. Logic alone can never free you from fear. Call upon your intellect, intuitive gut, imagination, emotions, faith in yourself, and your Soul when examining fear.

IDENTIFYING WHAT YOU FEAR MOST

Imagine yourself in a lab coat in front of a dissection table. Your task is to identify recurring fears and threats that make you anxious, insecure, or vulnerable. Identify each fear or worry as precisely as possible. You might have a dozen or more.

These questions may help you identify your fears:

1. What do I frequently worry, fret, or become fearful or anxious about when alone, especially before, during, or after sleeping?
2. Do they prevent me from sleeping or sleeping well?
3. Do I have a recurring frightening or troubling dream or dreams?
4. What fears, threats, or worries surface when I'm stressed?
5. Do I fear my strengths becoming compulsive and excessive? (These will be discussed as "crucibles" in a later chapter.)
6. Do fears, threats, or worries prevent or interrupt a peaceful state of mind, entertainment, family outings, or my favorite pastimes?
7. Is there a person(s) in my life whose predatory or bullying behaviors cause me to tiptoe around to avoid repercussions for me or my loved ones?
8. Do I worry so much about our community, society, nation, world, or the future that I cannot sustain happiness or optimism?

After defining each fear or threat, imagine placing it on the examination table before you. Pick up each one and examine it closely. Select the three to five fears that interfere most with your quality of life and peace of mind. Set them on the left side of the table to be dissected. Don't overthink your selections. You can revisit your choices and come back for the others. You are beginning a process that will become a skill over time. Trust your gut.

If imagining this is not easy, use a real table. Stand in front of it. Write the name of each significant fear on an index card or a separate sheet of paper in bold black marker. Capture facts or feelings related to each fear in another color or smaller writing. Visualizing in this way can be powerful and revealing.

Fear often causes behaviors we find embarrassing and uncomfortable. When we're fearful, we may do or say things that are "unlike us." Unacknowledged, unconscious fear takes charge of us without our permission as in these fear-revealing behaviors:

1. Dominating conversations by nonstop talking, interrupting, or dismissing others who try to speak instead of listening.

2. Experiencing excessive emotion or engaging in atypical behaviors triggered by certain people or subjects
3. Conversely, shutting down or withdrawing from conversation
4. Uncharacteristic, excessive alcohol consumption
5. Feeling trapped or desperate to escape a gathering or conversation
6. Engaging in catastrophizing, alarmism, and extreme exaggerations
7. Provoking or reacting to belligerent responses
8. Lying or excessive exaggerations about oneself and others
9. Routinely telling made-up events or stories to enhance one's image
10. Ascribing exaggerated importance to ourselves in others' lives, such as habitually referring to people we hardly know as close friends
11. Ascribing incorrect or wrongful motives to others when we cannot possibly know them.

Don't be embarrassed or judge your behaviors. Fear is the cause, and when triggered, it can undermine your thinking and dictate your actions. By exposing fears for what they are—fears, not reality—you can begin to form a healthier, more detached relationship with your fears.

DISCOVERING THE SOURCE OF FEAR

Many fears are misplaced or misguided because the real fear is disguised. To release fear's grip on us, we must resolve the root cause of a fear.

Return to the three to six fears you selected as most troubling and break each one into smaller parts to identify its origin. For example, if you feel afraid after reading a news article, consider why. Does your fear stem from feeling powerless to influence the changes necessary to address the world events you're reading about or watching?

General and specific fears

We can be anxious and insecure about specific perceived threats, but more often, anxiety and insecurity are an amalgam of undifferentiated fears. Vulnerability and anger are usually linked to a particular fear. We can lessen anxiety and insecurity by reducing or transforming individual fears. When you notice yourself feeling generalized anxiety, take a moment to name specific fears that may be fueling your anxiety.

Inverted pyramid

Fear can be conceptualized as an inverted pyramid.

Often, your top three to six fears or anxieties, such as career, relationships, life transitions, the state of the world, etc., will be too broad or generalized to resolve. Attempt to subdivide them into specific "subfears" with their own triangles to be further subdivided.

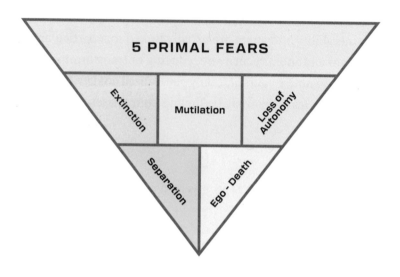

5 PRIMAL FEARS

Extinction

Mutilation

Loss of Autonomy

Separation

Ego - Death

Primal fears

The primal fears concept may be centuries old. It suggests that all fear originates from a few potent ones. Sigmund Freud introduced primal fears into modern psychology in a 1909 case study. More recently, Dr. Karl Albrecht significantly contributed to the primal fear construct by proposing five primal fears with complex emotional connections.[7]

One of Dr. Albrecht's primal fears is annihilation, which goes beyond the fear of death to ceasing to exist or never having existed, which is more terrifying than death. We have deeply ingrained attitudes and emotions that accept the certainty of death but strongly resist having never existed. The extinction fear discussed extensively in prior chapters may be rooted in the fear of annihilation.

A second primal fear is mutilation, going beyond what the word implies to a broader fear of invasion, loss of bodily or mental boundaries, or integrity. This is the primal fear many or most people have of animals and insects. It may be the source of fascination with real and fanciful human monsters such as serial killers and horror movie villains.

In non-physical form, it may be the fear behind loss of privacy, bullying, stalking, and what we feel when someone steals from us.

A third primal fear is loss of autonomy or being under someone or something else's control other than our own. We can relate to the fear of being in a claustrophobic place, unable to move or tied down and immobilized. But beyond physical forms, it may be the source of the fear of being trapped or stuck with no way out, whether in a job, a social situation, or even an unwanted vacation.

A fourth primal fear Dr. Albrecht calls separation, which extends to being rejected or abandoned or believing that we have been. Some Native American tribes did not practice capital punishment but instead used banishment from the tribe as an effective deterrent worse than death. Fear of separation and rejection is likely why some people today feel isolated, even though they are more connected than ever through smartphones, mass media, and social media.

The fifth primal fear is Ego-death, a broad term illustrating a threatened Ego's power over us. It includes humiliation, public persona devaluation, or threats to how we define ourselves. Technology has aggravated this fear for many when it should have reduced it. For example, everyday people can become celebrities through social media today. However, it also makes them more vulnerable to public opinion fickleness, where they can be rejected as fast as they are accepted or applauded.

Other psychologists have contributed variations to Dr. Albrecht's model, but it is helpful as is for our purposes of dissecting fear. Consider how the primal fear concept enhances the inverted triangle of fear, as shown on the next page.

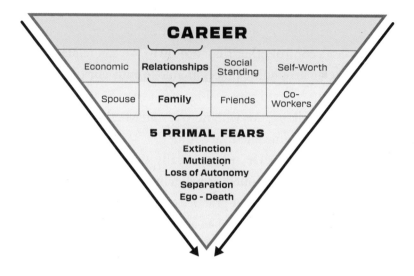

Attempt to connect the source of your three to six fears to one or more basic fears. This approach explains how seemingly insignificant threats become deeply rooted fears that provoke primitive reactions such as fight or flight. Be patient. Don't force a fit. You will know in your gut and heart when you have hit on the ultimate threat beneath your fear. There will be an emotional response or release. If no connection can be made, leave it for now and return to it later, or try another method.

Case Study: Nadine's Story

Nadine is a devoted mom of four children who is admired for being loving, patient, and tough when needed. As her children grew to young adulthood, she became increasingly fearful that they were unprepared for the fast-changing world. Nurturing became hovering and controlling, which made her kids miserable and rebellious. Fortunately, a wise friend asked her, "Who are you without your children?" Nadine was stunned to discover that she had no answer. Her misplaced fear for her children in a changing world was really about herself. She had unconsciously allowed motherhood to define her identity, similar to how people can be defined by their career, appearance, or social standing. Her real fear was abandonment and Ego-death, becoming a nonperson when her children grew up and left. Her wise friend recognized the archetypal empty nest syndrome and helped Nadine overcome that fear and embrace her next stage of life with calm courage. Archetypal fear of becoming a nonperson can arise during significant life transitions, including marriage, children, divorce, career change, and retirement.

SIZING UP FEAR

After isolating your fear(s) as clearly as possible, assess them unemotionally and critically using one or more of the methods below. This process usually reduces or eliminates fear(s) or clarifies precisely when and how to respond.

THREAT ASSESSMENT

Assessing fears puts them in perspective. Using a scale of 1 to 5, where 1 is the lesser threat and 5 is the most significant, give your fear a "score" In each of the following categories. A low-scoring threat may not warrant the fear attached to it.

- *Urgency*: (1) distant unknown time in the future, (2) within a decade, (3) within five years, (4) in the next few weeks or months, (5) imminent
- *Likelihood*: (1) unlikely to occur (5) virtually guaranteed to occur
- *Severity*: (1) unlikely to cause severe emotional, psychological, or physical harm or loss of quality of life; (5) severe, almost certain harm

- *Specificity*: (1) humanity, (2) a large population segment you identify with, (3) your community, (4) family and close friends, (5) you

The Ego can be persuasive that the world cannot survive without us to inflate our self-importance and vanity. We should all care about humanity but balance it with living our lives fully. We must accept that many things we would like to "fix" about the world are beyond our ability to influence. Also, the Ego's cognitive distortions warp our perception of severity and urgency. They cause us to demand impossible immediacy and seek perfection that does not exist. Our best course of action is to do what we can and trust that forces beyond us will do their part. Do your best and be satisfied that you have done all you can or should do.

FEAR BASED ON EGO COGNITIVE DISORDERS OR ILLUSIONS

The Ego evolved distortions and illusions to serve its hurry-up-and-survive-at-all-costs function. Illusions served humanity well for a time. Science, philosophy, and theology have explored and made some distortions and illusions obsolete over the last few centuries. For example, we now know that the Earth orbits the Sun and that the "vapors" identified by the ancients do not cause illness. Quantum physics is currently changing our understanding of reality.

Today, distortions and illusions are a source of fear that interferes with our ability to adapt to accelerating change. Recognizing and disavowing or moderating distortions, illusions, and their related fears allows us to embrace a more accurate view of ourselves and the world. We examine a few of the most widely held distortions and illusions below. Psychologists have identified dozens that are worthy of your time to research.

Ownership is stewardship

Much of our fear is rooted in losing what we work hard to acquire and protect, such as appearance, social status, power, money, family, spouse/partner, friends, love, health, and even our lives. These treasured aspects of ourselves and our lives often come to define us as they did Nadine. Ownership is an illusion; we are temporary stewards of everything and return it all as we age and die. We can aspire to and enjoy what we appear to "own," but in fact don't. This "ownership" becomes problematic when we:

1. Cannot separate who we are from what we possess
2. Do not believe we have value without what we possess
3. Believe life is meaningless without our possessions
4. Confuse ownership with life's meaning
5. Are willing to harm others to secure ownership of possessions
6. Define ownership as success
7. Believe we deserve ownership instead of feeling gratitude for what we have

Stewardship distinguishes between what we have and who we are.

Certainty isn't

As far as we know, nothing in life or nature is inevitable. Everything perpetually changes. We cling to the illusion of certainty to feel safe. Fear of change is a universal human fear because change forces us to confront the reality of uncertainty. Fear of change damages relationships, careers, and enjoyment of life. Consciously acknowledging uncertainty will need to become our new norm as change accelerates. A mindset that change is the only certainty will be essential for career success and peace of mind.

Straight lines aren't

Engineering, construction, and predictions rely heavily on solid foundations, baselines, and linear projections. Nothing is straight or static, even if it appears so. A rock seems unchanging but changes every moment. The arc of history is not one straight line forward but many wavy lines that intersect. Many fears are based on straight-line projections of a perceived condition or situation in its present form. Predicting tomorrow's conditions based on today's reality is misguided. The propensity demonstrates a lack of perspective and a dangerous blind spot. Creativity and innovation almost always solve known problems given time—necessity is in fact the mother of invention. The real danger lies in what we don't see coming. We are less likely to spot emerging threats if our attention is devoted to exaggerated or long-term threats we have time to address. Reject fears based on the illusions of linear certainty in your personal life and the world. Instead, consider how innovation could solve the problem and ask what more significant threat is being overlooked.

Death as illusion

"Shrinking away from death is something unhealthy and abnormal which robs the second half of life of its purpose."
—CARL JUNG

Most people fear death, but perhaps our fear derives from a misunderstanding of death itself. When we accept the law of energy conservation[8]—that energy can neither be created nor destroyed, only converted from one form to another—death as a final endpoint becomes an illusion. The human body contains bits of the universe through elements of the periodic table that exist inside us[9] and throughout nature. Instead of being permanently terminated at death, the body transforms into one or several different forms of energy.

Some ancient people believed they lived eternally through their descendants long before genetic science proved them correct. Genetic discoveries continue to expand our understanding of inherited attributes beyond physical characteristics and disease to include elements of personality. Artificial intelligence can create digital versions of humans that incorporate their experiences, memories, and temperament. Digital versions of humans can transfer knowledge and wisdom to future generations as tribal oral histories once did. Someday, we may all feel a moral obligation to leave an accessible record of our lives that descendants can know and learn from.

Medical science continually increases life expectancy, changing perspectives on aging and life itself, reflected in sayings like "Seventy is the new fifty; ninety is the new seventy." How will our attitudes about death change when the definition of death becomes more fluid, and lifespan expands significantly? Rather than fearing death as something that steals life from us, it is healthier to recognize that we waste life being fearful that is precious and needs to be lived well. **Death is temporal life's inevitable next act, not extinction but a natural cycle of completeness and renewal.**

STRENGTHENING YOUR SELF-SOOTHING SYSTEM AND COPING STRATEGIES

Hopefully, you have completed your first fear dissection: identifying your most challenging fears, discovering their source, assessing severity and urgency, and separating real threats from Ego distortions, illusions, exaggerations, and disguised threats. Congratulations, even if you only completed this process for one fear. Make fear dissection a habit. With practice, you'll be able to dissect fear in minutes. By reframing your fear, you will release yourself from its grip and take a big step toward your potential and expanded capacity for wisdom.

Dissecting fear to cut it down to size is a big first step to becoming calmly courageous. The second is engaging your innate self-soothing system. You already use it to some degree. Consciously and deliberately maximizing that self-soothing capability can become second nature enabling you to remain "calm under fire."

Self-soothing capacity varies widely from one person to the next. Some people, usually wise, are perpetually calm and impossible to rile. Others are overwhelmed by fear and vulnerability-driven emotions, making them prickly, volatile, and predatory when threatened or stressed. The majority of us are somewhere in between, probably in a bell-curve distribution. The strength of a person's self-soothing system is directly related to genetic personality predisposition, parental guidance, life experiences, and, most importantly, INME development. Let's explore powerful, intentional forms of self-soothing.

COLLECT YOURSELF

Fear, which derives from the Ego/Shadow, scatters, disorients, and sometimes triggers a fight-flight-freeze response, especially if it comes upon us unexpectedly. Engaging the INME allows you to tap into wisdom. When fear strikes, first use calming practices including deep breathing, ritual washing (face, arms, neck, and hands), taking a walk, going to a favorite place to reflect, connecting with nature to enter a state of Awe, or lavishing affection on a beloved pet.

If you are in a situation where you can't escape imminent verbal or physical threats, such as on a plane, in an automobile, or a scheduled meeting, politely excuse yourself: "I need a few moments to collect my thoughts and consider your point of view." The other person may bait you to escalate the conflict. Stand your ground and take that moment of solitude to collect yourself.

WORST-CASE SCENARIO

When facing a fear or threat, you have likely asked and answered, "What's the worst that can happen?" It's a simple and highly effective self-soothing method. Frequently, the answer eliminates the fear or relegates it to its rightful position.

SERENITY METHOD

The serenity prayer is a sage, succinct, actionable method of self-soothing. It reminds us to act on fears we can affect, and to accept the uncertainty of what we cannot change: "God, grant me the serenity to accept the things I cannot change, the courage to change the things I can, and the wisdom to know the difference."

This prayer has many applications in optimizing the performance of the psychological body. Among them is strengthening our ability to self-soothe when fearful or anxious.

TIMING

Sometimes, we face a threat or fear beyond our ability to address or soothe in the moment. All we can do is keep the fear in the back of our minds and bring it up again periodically to be examined. This process can be soothing if we patiently await clarity that will emerge in time and refrain from panicking, denying, or avoiding the fear.

BE A GOOD PARENT TO YOUR DEVELOPING INME

Parents play a vital role in developing children's INME self-soothing capacity by lovingly and patiently explaining a specific fear and how to self-soothe. Parents who shame children create fearful adults by teaching avoidance, suppression, and denial that strengthen fear. The healthy parenting model can be used in the same way we develop intimacy with ourselves. In this case, our INME is both parent and child and uses the healthy parenting approach to strengthen the innate self-soothing system.

ACKNOWLEDGE FEAR TO AN INTIMATE FRIEND OR PROFESSIONAL

Unacknowledged fear enslaves us, but revealing fear to ourselves, much less to others, can be daunting. Gather your courage, reach out to a trusted friend, partner, or professional such as a therapist, and honestly express your fear. Communicating fear does not resolve it, but by taking that leap, you initiate the change from fear avoidance or overindulgence to the self-soothing that will release you from fear and guide you toward calm courage.

Another excellent first step is writing about your fear. Draw its source triangle or use another method to move the fear from inside to outside. Read aloud what you know about the fear, including describing visualizations or drawings. Then, share your fear with a trusted confidant. You will be amazed at how well they understand and how empowered you feel afterward.

ENGAGE THE PSYCHOLOGICAL FUNCTIONS

Remember that you always have everything you need to create a healthier relationship with fear. The Ego and Shadow act as fear's messengers. They persist until your INME consciously accepts responsibility for resolving fear. The Soul and Conscience are the source of the self-soothing that offsets the Ego/Shadow's fear-filled messages. The Soul/Conscience and Ego/Shadow as opposing forces can't trump one another. The INME rises from the offsetting tension to mediate and resolve the fear.

THE KNOCKOUT PUNCH: TRANSFORMING FEAR INTO CALM COURAGE

"You gain strength, courage, and confidence by every
experience in which you stop to look fear in the face.
You are able to say to yourself, 'I have lived through this
horror. I can take the next thing that comes along.'
You must do the thing you think you cannot do."
—ELEANOR ROOSEVELT

The fear function has an energy reserve capable of providing almost superhuman strength and courage when needed. When a fear or threat is reduced or eliminated, dammed-up energy that was bound to it will attach to other fears, insecurities, or general anxiety unless consciously redirected by the INME into calm courage. Using dissection and self-soothing, you can effectively reduce or eliminate one fear at a time to construct and reinforce the calmly courageous INME.

There is another faster, more decisive way to overcome fear. In a single leap of consciousness, a life-changing aha! moment of clarity, you can become fearless when you were fearful a moment earlier. In that instant, you dare to overcome what once you feared and wished to avoid. You are no longer the person you were a moment earlier.

Seem impossible? Not if you have experienced other aha! moments. I have known many wise, calmly courageous people who described their moment in language remarkably similar to Eleanor Roosevelt's. Like all aha! moments, you cannot force them, but you can create conditions that make them more likely. One way is by answering one of life's unavoidable questions, *Do I have the strength to overcome whatever adversity life presents?* You can ask yourself this question, someone may ask you (as happened to me), or a situation may pose the question. Start pondering the question now so that you are ready.

FEAR CASE STUDIES

My story

Since birth, I have been blessed with high energy, curiosity, and a desire for adventure, undeterred by fears that troubled many others. Nevertheless, my mid-thirties were stressful and tested my resolve. I was a single dad with two athletic teenage sons whose events I rarely missed. I felt stretched as an executive leading hundreds of employees, a nighttime law school student, and a scavenger for time to socialize and continue my human potential studies. Unsurprisingly, I found myself worrying that I couldn't do it all and fearing that I would fail as a dad. Fortunately, I had a sage friend. As I fretted aloud one day, he said, "Ben, you seem very strong. Don't you think you can handle whatever life serves up?" It was as if a jolt of electricity hit me. "Yes, I can," rolled off my tongue straight from my gut with no bypass through my brain. I was confident and unshakable because the assurance came from my Soul through my INME. I never looked back, and fear has never gripped me again.

Fran's story

Fran was in her late forties. A talented woman with a good job, she was living a frequently miserable life because of phobias that kept her in and out of therapy for years. Her progress in overcoming her fears was minimal, and her condition had already ruined several promising relationships. One morning, she went for a walk by the river, heartsick over an embarrassing company event where her phobias were on display. Unexpectedly, she experienced a moment that changed her forever, obliterating her phobias and releasing her from the prison they had constructed inside her.

Fran told me the story of that moment five years later. She could not explain why or how her moment occurred. Perhaps those years of therapy coalesced, or she just got sick of letting her fears steal her freedom and her life. All she remembered was asking herself through her tears, *Why? Why?* Then suddenly, she resolved, *No more.* She made an appointment with her new therapist and made a firm commitment to face and overcome every one of her phobias under the therapist's supervision. When I met Fran, this woman previously controlled by a phobia of heights among other phobias, was on a challenging hike along twelve hundred-foot cliffs above the sea. She was calm, courageous, engaged to marry, happy, and free.

Fear can work for you through understanding and calm courage. **Fear owns you, or you own it. Only you can decide who's in charge.**

WISDOM STORY: WHEN A PREDATOR CONFRONTS CALM COURAGE

The storyteller is a healthcare expert and executive, business leader, entrepreneur, and savvy investor whom I have known and admired for over thirty years. His story is a test of courage and wisdom:

> Fortunately, I worked with ethical, principled professionals in trusting cultures throughout my decades-long career. Human foibles were occasional road bumps, but nothing I experienced was too difficult. The painful exception was a dishonest boss. From the beginning, red flags indicated a challenging personality, including unprofessional foul language, aggressive, in-your-face communications, and impatience with objections to his opinions.
>
> As an executive at a publicly traded company, my responsibilities included contributing to quarterly reports to the US Securities and Exchange Commission (SEC) containing detailed company performance information, including financial statements and management commentary. While preparing financial input for quarterly SEC 10Q reports, an associate noted that specific numbers differed from our submissions to the corporate

financial unit. When we asked why the numbers were different, the higher-ups said other activities in the firm caused the adjustments.

I was skeptical. Then all hell broke loose. The parent company ran short of cash, causing the CEO to request that I delay the next vendor payment and promised me it would be only once. The vendors accepted the one-off event. But then the CEO asked me to postpone the next payment, and I flatly refused. I had given my word there would be no more delays. The CEO applied extreme pressure to force me to comply, including insults and threatening to fire me, but I held firm. I expected to be fired, but before it happened, the CEO was fired for misdeeds, including falsifying independent board member signatures. A subsequent investigation discovered that the CEO and CFO had falsified financial records. Both were convicted and served time in federal prisons.

Such behavior is distressing, disappointing, and embarrassing for anyone in leadership. I felt responsible for the thousands of employees and to preserve a reputation for integrity, regardless of personal consequences. But it took an enormous emotional toll, which I accepted as the price of a clear conscience.

CHAPTER 10

Getting Out of Our Own Way

WE HUMANS HAVE AN UNCANNY KNACK for constructing barriers to our potential, mucking up relationships, and generally making fools of ourselves. There is a book there, I'm sure of it. Maybe a multi-volume set. For now, let's consider some of the more common, serious ways we self-sabotage. Using the practices, perspectives, and mindsets that follow, you will learn to get out of your own way and access your potential and wisdom faster and more easily.

ACCEPTING REALITY

ACCEPTANCE AS A PSYCHOLOGICAL LEARNING STATE

Acceptance is an effective but subtle learning state that can profoundly impact our lives and the lives of those we affect. Acceptance has a far-ranging impact because we tend to project dissatisfaction and intolerance of ourselves onto others, damaging relationships and careers.

Self-acceptance is a significant step toward your potential, but acceptance as a learning state goes further. Self-acceptance is directed inward. Acceptance turned outward can fundamentally shift our worldview to embrace the world and life as it is, learning to work with things as they are instead of fighting the natural laws of the universe, the world, and other people.

REALITY

We like to think we understand our reality. But as we've seen, reality is distorted and made illusory by our personal Ego and society's collective Ego (refer to Chapter 4 for details). Repressed fear, anxiety, and biases also distort reality.

Let's say that we succeed in seeing reality. Game over, right? No, we must then accept what we see instead of rejecting or fighting it. For example, how often do you look into a mirror and, momentarily or permanently, reject or criticize your appearance? Have you had a friend that others identified as dangerous or unhealthy, but you continued in the relationship until you experienced their destructive reality?

As psychologist and author Tara Brach described it, "Radical Acceptance is the willingness to experience ourselves and our lives as it is." She also said, "The acceptance of oneself is the essence of the whole moral problem and the epitome of a whole outlook on life."[1]

EGO DISTORTIONS AND ILLUSIONS AS THE OPPOSITES OF ACCEPTANCE

The Ego's cognitive distortions are opposites of acceptance. We must remove them and find the true neutral opposite to establish the tension needed for creative resolution.

Using perfection as an example, making things better is a neutral opposite to acceptance of the way things are. Making things better is achievable but seeking perfection is not and is an unachievable distortion. The Ego and Shadow resist acceptance and push you toward the illusion of perfection. *I could be happy, if only . . . I could get along with my family, if only . . .*

The pursuit of perfection keeps you dissatisfied, angry, and anxious, demanding that "things" change to make you contented. This never-ending search is the Ego at work, taunting you with perfect but unattainable outcomes. The Ego grows stronger when you are fearful, insecure, and anxious and does not want your fear resolved. Only tension between two genuine neutral opposites produces realistic creative solutions.

YOU ALREADY KNOW HOW

You have undoubtedly resolved issues through acceptance of a situation or a person. Perhaps you found a new way to get along with a family member or friend. Maybe your job or career became rewarding after you accepted it for what it was and saw it differently. Remembering when and how you were able to accept a challenging situation creates psychological muscle memory. Once you remember what acceptance felt like in the past, you can engage the Acceptance state consciously and deliberately at any time.

ACCEPTANCE PRACTICES

1. Reread Chapter 7 to refresh your memory for shifting to a desired learning state.
2. Mentally note the next time you feel frustration or dissatisfaction with yourself or someone else. It's easier to start with yourself. At the precise moment you feel it, stop. If possible, pause or excuse yourself and go to another room where you can be alone.
3. Revisit Chapter 4 and remember that much of what we dislike about ourselves and others are qualities of the Ego, which is neither you nor the other person. Your task with yourself and others is to bring out the INME.

4. Visualize or draw the polarities involved. Your anger or irritation will cause you to focus on the source of the dissatisfaction. Try to feel acceptance and dissatisfaction equally. Ask yourself: What would meaningful progress be?

As an example, I am the stereotypical absent-minded professor who gets wrapped up in an idea or discussion. When I do, it can cause me to become clumsy, forgetful, overly enthusiastic, and out of touch with my surroundings. For years, I struggled to accept this quality. I was embarrassed and occasionally humiliated by it. Eventually, I was able to frame the tension as:

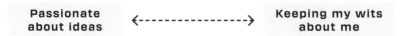

Both are admirable qualities, but I needed to find a new center. I adopted a practice of pausing to collect myself. When I sense myself overly caught up in an intense discussion, I pause to collect myself and say, "Is this making sense? Are you finding value in this discussion?" I also interject humor. If a discussion at dinner gets lively, I suggest to the guests around me that we remove any stemware or bottles of wine before my exuberant gestures bathe them in it. When writing, I set a timer on my phone to break my intense focus, drink water, collect myself, and become reacquainted with my surroundings.

5. Take an inventory of the qualities or behaviors that you struggle to accept about yourself. Prioritize them. Set a goal to address one per month or quarter, whatever you find reasonable.
6. Complete this same inventory when you experience dissatisfaction with others or some part of your life, such as your career or family.

Acceptance challenges and distortions of reality are serious. If you struggle intensely with distortions or your quality of life is being impacted, professional therapeutic techniques are extremely valuable and likely essential. For most people, turning dissatisfaction into action and progress creates positive change with an outsized effect.

RESPONSE TO ADVERSITY

ADVERSITY'S ESSENTIAL ROLE IN NATURE

Adversity in life is inevitable and essential in creation and the universe, as explained by the Law of Polarity. **Adversity is a teacher, not a punishing warden.** The frequency and intensity of adversity vary for reasons unknown. Understanding the role of resistance or adversity liberates us from feeling victimized or punished by what is a necessary natural force enabling you and creation to achieve potential.

ADVERSITY IS NEVER UNIFORM

One of life's unavoidable questions is: *Why do some suffer more than others?*

Life is more difficult for some humans, animals, trees, and flowers than others, partially explained by the Laws of Diversity and Nature's Rhythm. The Ego confronts adversity with resentment. It attempts to convince us that we should care only about ourselves instead of others because we are victims. After all, shouldn't adversity be equal for everyone? Wouldn't that be fair?

That is not reality. The Law of Diversity strengthens individuals and their species to survive. If all were equal, one disease might wipe out an entire species. Those with the greatest potential are often challenged the most because adversity brings out their best. A cultural archetype, especially in the United States, are stories of those who heroically rose from humble beginnings to become wildly successful philosophers, theologians, writers, engineers, physicians, businesspeople, inventors, and innovators. Another valuable lesson from those archetypal stories of success is that childhood and early adulthood adversity accelerated their potential and wisdom because they needed it to survive and achieve their dreams. The human species plays by the same rules as all living things. We are subject to a system of varying diversity, even if we wish otherwise.

ADVERSITY IS INEVITABLE— OUR RESPONSE IS A CHOICE

In *Man's Search for Meaning*, Viktor E. Frankl tells a remarkable story of Nazi concentration camp suffering and survival. His book illustrates the power of the human spirit to find meaning and love even when confronted with degradation and horror.[2] Thankfully, it is unlikely that any of us will endure anything remotely like Frankl's experience, but we can learn from how he and others responded in the direst circumstances.

You do not necessarily lack character and resilience if you feel beaten down by comparatively trivial challenges. Your pain is your pain. Your situation is your own. The lesson from the relative adversity of others is the importance of responding to adversity in ways that help rather than hurt you. You have no choice about encountering adversity. You choose how you respond to it. Learning from adversity is a decision to grow stronger and wiser and realize your potential.

ADVERSITY PRACTICES

Research and experience always bring me back to the same questions when I'm faced with adversity. Try starting with these the next time adversity arises and see if you find navigating it easier.

1. Is there a cause and effect I need to discover with this challenge, or did it occur through nature's version of randomized trials, and I made the list? Much of life's adversity is not personal but systemic selection. However, we all also have destructive life patterns that can cause some of the adversity we experience. Being aware of those patterns helps to avoid repeating them. Later in this chapter, we will look at how to recognize destructive patterns and prevent them from reoccurring.

2. How will I respond? What adaptive response will get me through the adversity as successfully and painlessly as possible? Resist the Ego's desire to convince you that you're a victim. Instead of resorting to anger and resentment, call on your INME to discover the lesson(s) you can learn from adversity.

CO-CREATORS

"If we are receptive and fortunate, we form symbiotic relationships with a handful of people in a lifetime where we nurture each other's health and growth like microscopic creatures."
—KENNETH PECK

Some people and relationships lift us to become our best. Others hold us back or destroy our potential, which will be the subject of Potentialist book three. Every relationship forever alters both parties. The environments we choose to be in shape us just like the people and relationships we choose. **Whom we spend time with and where we spend our time, in large part, make us who we are. Those decisions, made wisely throughout life, accelerate potential and wisdom.**

Ask yourself, "Do I spend time with those who encourage my best or worst, and if so, why?"

SYMBIOTIC RELATIONSHIPS

Kenneth, a longtime friend, likens co-creation to symbiosis in nature; each is a form of interaction between two organisms living in close physical association, ideally to the advantage of both. The psychological body benefits from symbiosis with those who have our best interests at heart. By listening to their experiences of potential and wisdom, we accelerate our own. As the best in us (INME) emerges, we naturally seek companionship with those who value our real self/INME and simultaneously avoid relationships with those who don't. Our relationships become a reflection of our best selves, able to nourish the collective Soul and make the world a little better.

CO-CREATORS DEFINED

A handful of the people we encounter in life affect us so profoundly that they become "co-creators" of our lives and we of theirs. Each of us influences who the other becomes. If the co-creation is positive, we cannot imagine life without the other. These are enduring, intimate relationships based solely on the joy of being together, with little or no other purpose or utility. Co-creators love unconditionally, warts and all, unaffected by time and distance. We all hopefully have one positive co-creator and are truly blessed if we have a few. Destructive co-creation also exists and can become an equally powerful co-dependency where the parties cannot imagine being apart even as they destroy each other.

INTERIM CO-CREATORS

Some co-creators influence us during a defined period yet have a long-lasting impact on our lives. Mentors and teachers, for example, may become co-creators yet move along once their role in our lives is complete. These interim co-creators promote our growth by engaging our INME and advising us on navigating life's inevitable adversities. Trained therapists can be particularly effective interim co-creators. Their skills and knowledge align with your goal of achieving your potential through your INME. Therapists understand how to maintain the objective distance and neutrality needed to see your reality clearly and promote your INME's growth rather than making your decisions for you. These are complex, challenging skills to learn and master. Therapists train for them, and they are also sometimes found in wise laypeople.

INVOLUNTARY RELATIONSHIPS

Most of us have at least one unhealthy relationship that we feel obligated to continue. No relationship is truly involuntary, but your sense of obligation may make it feel that way. Such relationships can be family, co-workers, professional, or long-term associations. These involuntary relationships can be destructive, causing damage that takes years to overcome. To limit their harmful effects, boundaries are required. If the other person refuses to stay within the boundaries, or you find that you cannot enforce them, you may have no alternative but to limit time with the other person or exit the relationship.

IDENTIFYING WISE CO-CREATORS

You can learn to identify and recruit wise, positive co-creators. Reading the works of wise people or having one as a co-creator helps you to recognize wisdom in others. These are some of the qualities or characteristics to look for when considering someone as a co-creator:

- Has the capacity for intimacy
- Emits a sense of calm, courage, and confidence
- Actively engages in different forms of learning
- Goes through life seemingly at ease

- Laughs and enjoys life but not at others' expense
- Always seem to have your best interest at heart
- Does the right thing, often recognizing it when you do not
- Philosophically grounded with a broader, deeper perspective
- Optimistic but realistic about challenges
- Asks excellent questions that challenge you
- Exhibits humility

CO-CREATOR PRACTICES

Being deliberate and objective when choosing relationships and exiting those you cannot improve is an important skill of living wisely and well. These questions help you assess the value of your relationships:

1. Does the relationship bring out the best in both parties?
2. Does the relationship exist solely for the joy of being together, or does some personal advantage, social purpose, or other utility take precedence? It is normal to have some utility, but it should never dominate a relationship.
3. Do both parties want the best for the other, even if it would end their relationship?
4. Are both parties growing as a result of the relationship?
5. Are both parties equally valued, even if unequal in maturity, experience, or social status?
6. Are you continuing an involuntary relationship(s) that fails the values test in items 1 through 5 above? If so, why? What boundaries can you construct, or changes can you make to improve its impact on you and your life?

FAMILY CO-CREATION

At its best, family is a manifestation of intimacy, unconditional love, and belonging, like a protective membrane between you and the rest of the world from birth to death. Family connections are unique. They often have a prewired intimacy based on shared DNA and experiences. As young family members mature, a healthy tension develops between the family collective and younger family members emerging as fully formed independent adults.

Most parents attempt a balanced response during this transition, respecting the normal tensions of the archetypal baby bird leaving the nest. This change in the family dynamic challenges the child, the parents, and the family collective. As the maturing child struggles to be their own person, the family must adapt and accept the adult they are becoming. If any family member is Ego-dominated, they may discount the value of family and avoid family intimacy in an attempt to control their environment. Wiser family members understand that being an individual within a collective is an unavoidable life skill required by family and society.

Family dynamics intensify emotions. We expect relatives to be more forgiving than the world, but often the reverse is true. Family co-creation improves when we:

1. Focus on improving our behavior, unconditional love, and commitment instead of highlighting other family members' shortcomings.
2. Spend time with people who have no family to better appreciate our own.
3. Ask ourselves, who would be there for us if we were sick, broke, or ill? It is most often family.
4. Families need time together to deepen intimacy, just like individuals in a relationship. Ask yourself if you are investing enough of your best self in your family.

INGRATITUDE

"I would maintain that thanks are the highest form of thought
and that gratitude is happiness doubled by wonder."
—G. K. CHESTERTON

Chapter 7, "Learning States," identified Gratitude as a voluntary state with distinct feelings and qualities we can enjoy at any time. **It is impossible to become wise without gratitude.** Repeatedly choosing ingratitude has serious consequences; people are likely to avoid, distrust, and judge your unhealthy attitudes. Philosopher David Hume defined ingratitude as "the most horrible and unnatural crime that a person is capable of committing."[3] Philosopher Immanuel Kant described ingratitude as "the essence of vileness."[4] Scientific research links ingratitude to unhealthiness. Studies over the past two decades have demonstrated that saying thank you and, more importantly, feeling thankful, has real and lasting effects on your overall well-being.[5]

Gratitude is healthy and attractive to others, but more importantly, it is a complex, essential part of being human. Neel Burton, MD, described gratitude as "not a technique or a stratagem, but a complex and refined moral disposition." It has poetically been defined as "the memory of the heart" (Jean Massieu), "the moral memory of mankind" (Georg Simmel), and "the queen of the virtues" (Cicero).[6] It is the recognition that good can come from parts of life that are outside us and beyond our control—other people, nature, or a higher power—and that owe little or nothing to us. Gratitude differs from appreciation, which is the recognition and enjoyment of the positive qualities of a person or thing. The dimension of awe, wonder, profundity, or humility is the essence of gratitude.

Gratitude is essential to the art of living well, and its absence causes pain and heartache but can be remedied. If you, a friend, or a loved one suffers chronic ingratitude, please seek professional help to address it. Many people, including me, choose to begin and end their days in moments of Gratitude. If you already practice Gratitude, consider helping others who do not appreciate its joyful experience, health benefits, and importance to living wisely and well.

MEMORY AND THE PAST

"The absence of memory or the inability to recall memories properly in an emotional context leads to dysfunction, but paradoxically, memories that generate too much emotion can be equally disabling."[7]

How should we accept what happens to us without allowing it to define us or limit the quality of our life and future?

Phil Knight, founder of Nike, describes memory's fallacy in his excellent 2016 autobiography, *Shoe Dog*.[8] He drafted his story of founding Nike and shared it with his original management team. He was shocked at how decidedly different each person's memories of the events were from his own and from each other. If you've discussed childhood memories with siblings, friends, or workplace memories with former co-workers, you can relate to Phil Knight's experience.

Legal professionals discount the value of unverified eyewitness testimony and clients' memories without secondary source confirmation. Law enforcement professionals consider eyewitness testimony to be the beginning of a long fact-finding process, not the end. Many people remain tragically defined by memories, many inaccurate, that limit their potential and quality of life. Psychological identification with remembered event(s) is a common cause of continued difficulty in life. Generally, we overestimate the accuracy of our memories, but some memories are deeply imprinted, even if exaggerated or distorted. My life has been relatively trauma-free, but I have known many people who have learned to separate traumatic memories from who they are.

Separating memory from who we are, the INME, is liberating. Instead of deciding that what happened or what you remember defines you, adopt a different mindset.

A friend is a remarkably successful, self-made woman despite a horrendous childhood. Her parents loved her but succumbed to their demons. At ten years old she assumed the adult role in the family that neither parent could fill. She had to overcome many trials, including teasing and bullying because of her parent's notoriety. After years of therapy, she courageously chose to speak of it with her parents, who by then had their demons in check and functioned as responsible adults. It was a painful but wholly beneficial confrontation. Fifteen years later, they have a beautiful relationship, and she is trauma-free. Lovingly but firmly explaining her experience and requirements allowed her to exorcise their demons from who she is.

"I am not what happened to me; I am what I choose to become."

—CARL JUNG

Jung's guidance underscores this same healing psychological principle, which I witnessed when a delightful, cherubic twenty-year-old man was involved in a horrific automobile accident in which others died through no fault of his own. Devastated by PTSD and survivor guilt, he suffered from nightmares and debilitating memories. Everyone who knew him became deeply concerned for his welfare. With the support of his loving family and the guidance of a therapist with adolescent trauma expertise, he was able to free himself from the trauma after several years.

Not all life-defining memories arise from trauma. Some people cling to idyllic memories so tightly that it limits their growth and potential. You've likely seen it manifest as the "yesterday's hero archetype" at athletic events where a parent or loyal fan attempts to relive inflated sports memories vicariously. One otherwise talented and mature woman had a series of failed relationships with men because hosting girlfriend events was her highest life priority. She even designed a home to be like a hotel to accommodate her gatherings. Even after realizing that her persistent focus on these events had become unhealthy, she failed to moderate her obsession, which was tied to what she believed to be the pinnacle of her life being the popular ringleader at a female-only prep school.

Someone compelled to duplicate idyllic memories might ask

themselves: *Was there a period I remember as the best time of my life? Do you try to repeat it? Are the memories possible to duplicate? How much priority do you give these efforts?*

Practices for Letting Go of Life-Defining Memories

1. Don't suffer in silence if you have memories that hold you back, limit your quality of life or future, or that you cannot separate from who you are. Therapists excel in this field. Find one you trust to help unload this baggage.
2. Don't judge yourself for being unable to unload it on your own or minimize your pain compared to others. Accept what you feel and act on it.
3. Begin the process on your own by accepting memories as only partially reliable imprints, not reality, and as something distinct, something that happened to you, not who you are or what you want to be.

PATTERNS

"Insanity is doing the same thing over and over
again and expecting different results."
—ALBERT EINSTEIN

Friends and audiences frequently ask me, "Why is it that we continue to repeat mistakes?" Everyone makes unwise decisions and often repeats them in recurring patterns. Many of your most damaging and destructive life experiences are patterns. Many people seek therapy to escape them. Wise people learn to detect and address repeated mistakes to advance their potential and wisdom. Let's look at why patterns occur and how you can identify and change behavior to break destructive patterns.

Life challenges us to grow and become wise. When we fail to learn from our missteps, we often face similar challenges again, sometimes clearly, other times disguised. Parents teach their offspring by repeating tasks with slight variations until they perform correctly; we see this play out in humans and animals alike. Education uses the same approach. It's one way that nature nudges us (and all living things) toward our potential.

Pioneering psychologist Sigmund Freud identified a repetition compulsion in some people. Patterns aren't recognized for a variety of reasons. Some fail to see the pattern; each situation seems unique even if a slight variation of previous troublesome situations. Others recognize patterns but don't know how to identify the cause. Others see the pattern and the cause but don't know how to address the causes. Unaddressed causes can continue for a lifetime with subtle variations and increasing consequences that can bring you to your knees. Some recognize the pattern, the cause, and a solution but apply a coping mechanism to a new situation that previously worked but proves inadequate in the new situation. The trick is experimenting with what works and doesn't by being resourceful and flexible to address the cause or adopting a new coping technique if the cause can't be eliminated. You need not struggle for years or a lifetime with a pattern. Look at it as an opportunity to learn what you missed before. Pay close attention when something goes awry to look for a pattern and cause to address.

Let's consider four types of patterns: (1) societal archetypes, (2) crucibles, (3) personality predispositions, and (4) other causes.

Social archetypes are timeless patterns of human behavior. Examples include the prodigal son, the wealthy miser, the bully predator, the harridan, and the resentful, less-favored brother, as in the biblical Cain and Abel. Much of life is played out unconsciously and unwittingly through archetypes. If you know you are in one, you can change the outcome. If you don't, you'll be an actor in a predictable plot.

A crucible is a pattern of repeated persistent, destructive mistakes that many people experience. Thankfully, they are predictable, making cause identification and cure easier. The *Merriam-Webster Dictionary* defines a crucible as "a severe test" or "a place or situation in which concentrated forces interact to cause or influence change or development."[9]

Crucible in the context of patterns or repeated mistakes fits both definitions. Crucibles are destructive excesses of our most positive attributes since birth. This phenomenon has long been colloquially expressed as: "Our greatest strengths become our weaknesses." An example is someone who is so kind and caring that they fail to set boundaries, leading to abuse of their kindness. Another example is a tenacious person who overcomes obstacles but sometimes does so by pushing themselves to dangerous levels of mental and physical exhaustion. A prior chapter described the importance of knowing the "constant you" and your dominant traits before age six because they will likely remain your dominant traits and talents throughout life. They will also, in excess, be the source of your most destructive patterns. Everyone has three to five gnarly crucibles that show up repeatedly. You will save yourself significant pain and heartache by addressing these proactively. You can help emerging young adults to watch out for theirs.

Personality predispositions are like crucibles: excessive tendencies tied to a specific personality trait, but typically not so deeply rooted or destructive. Every personality trait is a tension of opposites whose polarities represent the aspirations and interests of the individual and the collective. The psychological body mediates opposites through the repeated consequences of excess.

Other patterns of repeated mistakes

Patterns of repeated mistakes are sometimes unique to our personality or life experiences rather than social archetypes, crucibles, or personality dispositions. You address them similarly, looking for the cause, but know that the patterns will be more unique and less predictable.

DISCOVERING YOUR PATTERNS

It is difficult to recognize your repeated patterns, but easy to observe them in others, as you mutter, "Will they *never* learn?" We should be asking ourselves, *Why do I* never *seem to learn?* Friends, family, co-workers, and others see your patterns better than you do. Those who have known you from birth can be especially helpful with crucibles, pinpointing them so we can make changes to avoid repeating a destructive pattern in the future.

1. List the patterns that repeat in your life, causing inconvenience, distress, or pain.
2. Ask friends and family if they recognize the patterns that you may have missed and, if so, to discuss them with you. Ask them if they identify patterns other than the one(s) you described to them.
3. Attempt to classify the patterns into social archetypes, crucibles, personality predispositions, and unique personal patterns.
4. What personality trait (usually a crucible or predisposed trait excess) could be causing your patterns?

THE UNTRUE STORY

Are the stories you use to describe yourself true?

It is normal to develop useful stories about yourself, such as verbal and physical bios or resumes, and those that summarily add color and insight to who you are or wish you were, what you like or excel at, and what you don't. These stories become your public and private identities—your personas. Carl Jung defined personas as "the personality that an individual projects to others, as differentiated from the authentic self" (INME).[10]

We all have at least two personas, and usually many more that are derivations of our INME, adjusted for the audiences or roles we play in life. Generally, personas fall into two categories, "private" and "public," with multiple personas or story versions in each. Few of us are taught the importance of personas to self-knowledge and self-acceptance or how to craft healthy personas instead of destructive ones. These self-defining stories or personas need your active attention. Learning how to construct or modify them allows you to use them to achieve your potential and become wise, and to communicate more clearly to others what you would like them to know about you.

PERSONA FUNCTIONS

Personas protect you from revealing vulnerabilities to those you do not know well or trust. They also provide social efficiency by setting generally accepted boundaries for self-description, making it unexpected and unnecessary to reveal more than is appropriate for a particular role or situation. When generally accepted boundaries are crossed, audience discomfort is evident. You've probably witnessed social gatherings where someone tells their life story in excruciating detail to people they are unlikely to see ever again. Watch those listening squirm.

THE PROBLEM OF UNTRUE STORIES

If your INME is present in consciousness, even if not yet dominant, your personas likely were generated by the INME and are genuine, if partial, projections. Each persona is tailored to a specific role or audience, like how you select clothes for a particular setting. Personas rooted in the true self INME do not create anxiety or fear of being called out, because they are the genuine you, purposely incomplete.

When the Ego dominates consciousness, it creates untrue stories that inhibit your growth and become problematic in your life and career. The Ego's prime mission of defense creates personas that are often deceptive, self-serving, and often embarrassingly fictional. Blind to the true self

INME, the Ego invents stories it thinks will make you safe or advance you in the world. Your public story becomes what your Ego thinks others or society wish to see or defies their expectations if the Ego deems rebellion to be safer. Ego personas are communicated verbally and nonverbally through behaviors, material possessions, dress, and even choice of name. Significant gaps develop over time between your INME and Ego personas.

The Ego also corrupts private personas. Instead of creating peace and comfort, the Ego crafts an untrue story for those closest to us, and even ourselves, in a misguided search for self-acceptance that never materializes because the story fails to match our true self INME.

A THREATENED EGO

Common causes of untrue stories include a lack of self-knowledge, self-acceptance, and feelings of inadequacy. The threatened Ego concocts an identity it believes makes it safer or more important, usually riddled with exaggerations or fabrications instead of your real personality, experiences, and accomplishments. In other cases, the Ego's untrue stories justify self-perceived shortcomings through a fictional unhappy childhood, lousy parents, and other life events concocted to make you sound like a victim. Other untrue stories dwell on unresolved wounds, some real, others made up, that have yet to heal. Still others try to justify or hide behavior or past acts.

IDENTIFICATION

False, self-defining stories can result from equating your career, level of fame, or societal position with who you are. Role identification indicates a lack of confidence in your reality compared to the role you hold. For example, someone may psychologically identify as Pete Smith, CEO, instead of Pete Smith, Iowa farm boy, or as wealthy socialite Mary Carter, not Mary Carter, mother of three. This is a common and unfortunate social norm, especially in the United States. I became intrigued by people asking five minutes after meeting me, "What do you do?" I began to have fun with it, responding with answers such as: "At what time of day? With what?" and other silliness. It was a humorous way to make a point of how we attempt to classify unique people into social strata instead of asking questions that would really inform, such as these: What's your story? What is your favorite activity (memory, hobby, etc.)? Who are your heroes? And so on. **What we do can change unexpectedly, but who we are is forever, so differentiating them is essential.**

CONSEQUENCES

Untrue stories are common today and often treated as harmless puffery instead of lies. However, they inhibit personal growth and may cause permanent brand damage if the gap between your INME and stories are widely divergent. Severe anxiety and vulnerability develop from fear of discovery if they diverge so far as to make us a fraud or if we believe we are one. Vulnerability leads to unpleasant visits from the Shadow that can make life miserable. Deviations are more observable by others than generally believed. We appear "uncomfortable in our own skin." It becomes increasingly difficult to keep stories straight. People with high emotional IQs and seasoned leaders are skilled at detecting a lack of genuineness, leading to relationship and career issues. As the truth emerges, careers and lives are destroyed, as witnessed in the tragic downfalls of politicians and other high-profile celebrities.

TAKING ACTION

Seeing and accepting undistorted reality is emphasized throughout this book. It is vital that the personas you project, your stories, are an accurate projection of your INME. It will become essential in the increasing transparency of the New Reality, where anyone can be researched.

Practices

1. Chapter 4 explained how to engage the Soul and INME so the Ego does not dominate your consciousness. You can engage your INME to replace your untrue stories with reality even if you are just learning to relate to your INME.

2. Accept who you are and confide your reality to people you trust. This opens the door to discovering your true self INME that will eventually crowd out the Ego fiction entirely.

3. Pay close attention to when and how you tell your story in the future. Reflect on the stories you tell now and how they can be rewritten to reflect your reality. Don't be embarrassed. Those stories were developed at one level of consciousness. You are now at a higher level. Rewrite the stories in your mind and on paper. Say less. Gaining confidence that less is better will allow you to build a solid foundation of reality incrementally. Gradually replace fiction or exaggerations with reality. Practice telling your new authentic story. You'll be pleasantly surprised by how you feel and how others respond.

As you become wiser and realize more of your potential, return to this chapter and any others you find helpful, at least once each quarter of the coming year. If that seems too much, back off to two or three per year. Sometimes, a simple change in perspective, or one new practice used regularly, is all you need to get out of your own way.

WISDOM STORY: FORGIVING THROUGH GRATITUDE

Hao Harrison is a remarkable woman in her early eighties. She grew up in Vietnam during the war, married, and had five children by the age of twenty-nine. In the war's chaotic closing days, she escaped to Indiana with her children, thanks to the heroic efforts of her US Navy officer brother-in-law. Hao spoke no English and struggled to create a life for herself and her kids in the United States. All became highly successful, blessing her with eight grandchildren. I met Hao in 1976 as a co-worker at Blue Cross and Blue Shield of Indiana. We became close friends. Hao has many stories of tragedies endured and great joy. This is a favorite of mine.

When Hao was thirteen, her father was severely injured by a cut to his arm at age forty-six. He was the only support for his family of nine. Surgery, a hospital stay, and a long recovery were required to save his arm. His savings and assets were inadequate to support his family during the recovery. A more prosperous family nearby offered to loan him six hundred dollars if Hao, as his oldest child living at home, worked as their live-in domestic servant for a year.

There was a catch; Hao would not be allowed to visit her nearby home and loved ones until the debt was paid. The family she served didn't beat or starve Hao but showed little empathy or kindness. She slept each night on a rice mat on the floor and was heartbroken each day seeing the home and siblings she could not visit in the distance. But she did her job well and always gave her best. She completed her servitude after a year and was reunited with her family.

Fast-forward to 1993. Family members from Hao's time as an indentured servant immigrating to Indiana sought her help. Remarkably, Hao shared her time, food, few possessions, and kindness with people she could have justifiably ignored. When I asked Hao in 1993 why she chose kindness over resentment, she said they saved her dad's arm and, through that act, his entire family. She expressed no bitterness over her treatment. Instead, through gratitude for the opportunity to help her father and family, she found forgiveness for unkindness. I found her wisdom inspiring and hope you do as well.

CHAPTER 11

Your Potential and Wisdom

THE DECISION TO PURSUE YOUR POTENTIAL AND WISDOM is personal and private. How you navigate this choice will define your life. It is the most important decision you will ever make.

I made my decision as a young adult. I know what it did for my life and loved ones and what it can do for you. I have devoted my life to studying wisdom, often firsthand, in the company of wise people. I've observed the qualities that made them wise and the paths and decisions that brought them to their place in life. They were universally driven to do their best while improving the lives of others, which was reflected in their decisions. They lived very well as a result. Their lives exceeded their dreams, and were abundant in purpose, meaning, and intimacy, as mine has been. I want that for you.

Your decision to pursue your potential and wisdom is more crucial than it was for prior generations. Through no fault of your own, you are unprepared for the unprecedented times ahead. For those before you, the pursuit of potential and wisdom was optional; for you, I see no rational alternative that won't result in unnecessary suffering. The unavoidable question you must answer for yourself is: *Will I adapt to the most rapid and extensive changes in history by becoming my best and wisest, or will I do nothing and endure the suffering that results?*

This chapter will help you answer that question.

TOUGH DECISIONS IN A TURBULENT TIME

If unprepared to be wise, how will you answer unavoidable questions that will take on unprecedented significance in the years and decades to come? These questions include:

- *What does it mean to be human?*
- *Will working and living with humanlike automation emphasize or blur my human distinctiveness?*
- *How do I integrate with automation like AI and robotics without losing my humanity?*
- *What security, backup, and controls should I insist on as my life becomes increasingly dependent upon ever more intelligent and competent automation?*
- *How do I prepare for career opportunities as automation assumes the grunt work and frees me for higher-value work that is creative, problem-solving, and involves complex relationships with humans and humanlike automation?*

You will work, live, and perhaps raise a family or assist the one you have through the transition from Industrial Age practices, structures, and mindsets to the Age of Expanded Human Potential. The Industrial Age unfolded over roughly 250 years; the Age of Expanded Human Potential will likely develop within an unfathomably rapid thirty years. *Will you navigate rapid change to seize opportunities or be frozen by fear and blinded by pessimism?*

There is abundant evidence of declining confidence in institutions of government, education, mass media, healthcare, religion, and some large corporations. This decline is due to democratization, deinstitutionalization, and disruption. It will likely worsen before new revitalized institutions emerge. You can choose to self-sabotage with negativity or innovate and discover new possibilities for you and your life in these rapidly changing times. *If institutions continue faltering, how, when, and on what matters will you fill the void they leave, protect your interests, and participate in reshaping them?*

Social and political divisions herald the transitional time ahead. Conflict and division in society and among families and friends are troubling. No unifying visionary leaders have emerged, but we don't know what we can do as individuals. Chapter 3, "The Law of Polarity," explained polarization and tension as nature's birthing process for creative resolutions, as wise people have long understood. You can integrate these fundamental truths and prosper or resist them and endure the consequences. *Will you lead wisely to heal polarizing issues in your orbit or stay silent and hope someone else leads?*

Population decline will be a defining issue in developed countries through the mid-twenty-first century and worldwide after 2160. No country has a plan for maintaining a safety net for the elderly in an aging population with fewer young people. This is not an issue you can ignore. You must have an adaptive strategy for you and your loved ones, such as optimizing your health, accumulating financial reserves, becoming your own competent health advocate, and working longer. *Will you develop and execute an adaptive strategy for a longer life in an aging society, or will you trust that the government or someone else will take care of you and yours?*

You have the opportunity to live better and longer with expanding, unprecedented choices for bespoke health, careers, products, services, learning, and lifestyles. The choices will often be unclear. How you handle the ambiguity will shape your life. Or you can let others make these critical decisions for you and risk the outcome. Ask yourself: *How will my loved ones and I assess the choices and consequences to make the wisest possible decisions in a personalized world of expanding choices, individual initiative and productivity, and self-discovery?*

These unavoidable questions are not intended to frighten you but are practical examples of questions requiring wise decisions that you already face or will face very soon in the New Reality. Many others are presently unknown but coming. No one is going to answer these questions and make decisions for you with the same care you would give to them. If you are unwise and unprepared, your Ego, Shadow, and the fears and emotions they generate will make these decisions. You won't like the result. You can't avoid the times or the decisions you will face, but you can be as prepared as possible, guided by the INME and Soul, to make the decisions fearlessly and wisely. Your quality of life and those of your loved ones in the emerging New Reality will be what you, your peers, children, and grandchildren make of it.

VISUALIZE YOURSELF OPERATING AT YOUR POTENTIAL WISELY

Carl Jung's "active imagination," is a very old, proven, powerful skill that is easier to remember as "visualize-to-realize." **Active Imagination or visualize-to-realize is concentrated practice to learn and become what we imagine or visualize.** It is an important adaptive skill for the New Reality because the pace will be too fast and consequential for you to keep up unless you visualize where change is headed and how it affects you. It is a skill enhanced by a developed INME that can access your Soul for insight, inspiration, and creativity.

Active Imagination or *Visualize to Realize* has been discussed and encouraged throughout this book. It is a skill you've used hundreds of times to learn and improve in sports, dancing or acting, games, business, and planning social or family events. In Chapter 4 and later chapters, we explored the characteristics of the Ego, Shadow, Soul, Conscience, and INME. The descriptions were detailed enough for you to recognize and work with each function by actively imagining it. In Chapter 7, "Learning States," we described each state thoroughly so that you could recognize when you're in each state. Then, using the descriptions, reenter states at will, switch, and multitask learning states. Now you will visualize yourself as wise in order to realize that wisdom over time.

Many characteristics or qualities are consistent among wise individuals, as demonstrated in an article by Dr. Andreas Fischer that compares the common qualities and teachings of Socrates, Jesus, Confucius, and Buddha.[1] Below is a composite profile of many qualities of wise people drawn from literature and the wise people I have known. No person possesses all, and you need not aspire to all the qualities to become wise. The purpose is to create a general profile in your mind so that you can easily recognize wise qualities in others and select those you aspire to.

Your active imagination/visualize-to-realize skill test is to construct and achieve your personalized wisdom profile. Identify the wisdom qualities you already possess, including those below. Then, identify the qualities to which you aspire. Together, the qualities you have and those you desire define your aspirational best self. Write down or record your aspirational vision. Read or use voice playback to frequently review different parts of the profile in manageable segments, such as five minutes a few times a week at bedtime. This imprints the profile in consciousness over time. Passionately desire to become that person. Focus on mastering one or two qualities at a time. Consciously and unconsciously, you will begin to emulate the characteristics of the wise through the INME and see opportunities to test them out. Being around people who lift and inspire you, such as mentors, accelerates the process. In some ways, it is similar to learning a new language. You immerse yourself in wisdom to become wise.

Don't be surprised by negative thoughts as you read these qualities or start to use active imagination. Unwanted thoughts may pop into your head, such as *I can't do this, this isn't me, it's beyond my ability,* or *no one can be like this.* The doubts come from the Ego and Shadow, seeking to retain domination of your consciousness, everything you think and do, through tricks like perfectionism, self-doubt, self-loathing, and fear. You may also experience resistance from the collective Ego in person or in things you read or watch. It uses intelligent but Ego-dominated mouthpieces to create doubt about wisdom, wise people, and the existence of free will.

Recognizing negative thoughts as products of psychological

functions—the Ego and Shadow—and not your own is an important learning opportunity. Don't repress or shy away from the negative thoughts. Instead, express in response what you feel, such as:

- "No one is perfect, even the wise."
- "I can be wise. Wisdom is inside me."
- "Pursuing my potential and wisdom is an honorable pursuit. I'm unafraid of criticism."
- "This is my decision, not yours, Ego."

If negative thoughts distract you so intensely or frequently that you cannot concentrate, your Ego "has you" and is controlling your consciousness. Set the practice aside, select a few practices to quiet your Ego, connect with your Soul, and begin again or return to the reading later. Don't give up. This is your struggle for psychological freedom.

The characteristics of wisdom below are grouped into seven categories. Feel free to organize them differently and add other characteristics to make them as useful and relatable as possible. Take your time going through them. Read and reflect from a calm, quiet place inside. Savor and visualize them as your own. Plan to repeat the imprinting practices over the coming weeks, months, and years. You will be surprised by how many characteristics and qualities your INME effortlessly incorporates over time. **We can be only what we imagine, and what we imagine, we can exceed.**

SEARCHES FOR TRUTH, REALITY, AND PERSPECTIVE

- Eager and able to struggle with life's greatest questions (existential and unavoidable)
- Seeks the broadest perspective and deepest essence in pursuit of truth, reality, or what might be at work beyond seemingly narrow issues, everyday matters, events, or arguments, often discovering what others miss

- Seeks reality in the outer world and the inner world of the psychological body
- Integrates an understanding of nature's laws, patterns, archetypes, and symbols into daily life, problem-solving, and decision-making
- Recognizes, appreciates, and aligns with nature's rhythms, including ups and downs, twists and turns, and the expected and unexpected
- Loves life as it is while always taking action to improve it, starting with themselves, loved ones, and others in their orbit
- Observes the parallels between the natural world and human behavior and vice versa (the macrocosm-microcosm theory)
- Understands and commits to overcoming Ego illusions, fictions, and cognitive disorders distorting reality in themselves, others, and the collective
- Senses and appreciates beauty in life that others may not recognize or appreciate
- Accepts pain and tragedy as part of temporal life, recognizing that evolution and universal expansion improve the human condition over time

BALANCED, PRAGMATIC OPTIMIST

- Lives a full meaningful life with one foot planted firmly in the temporal world and the other in the eternal, honoring the Ego/Shadow's defensive functions and the Soul/Conscience's aspirations for the whole
- Understands the law of polarity and uses it extensively in conflict resolution, decision-making, relationships, and personal growth
- Pursues balance in life and personality, avoiding unhealthy extremes while enjoying individual differences in self and others
- Gives balanced attention to the psychological and physical bodies, increasing priority to the psychological with age

- Embodies humility despite obvious natural talents and an impressive track record of accomplishment, but recognizes, honors, and eagerly applies their unique talents
- Accepts their best effort instead of chasing illusory perfection
- Remains centered, balanced, solid, never erratic or volatile
- Honors and balances their needs with those of others
- Experiences success in the temporal world while remaining spiritually connected
- Believes in people and humanity and is optimistic about the future, recognizing and accepting adversity without guarantees

INTIMATE RELATIONSHIPS

- Understands intimacy's foundational role in love, empathy, and relationships
- Sustains intimate, longstanding, rewarding relationships with a variety of people and the Soul
- Accepts, appreciates, and desires intimate relationships but chooses them carefully
- Seeks intimacy without the Ego's need for utility or purpose
- Allows relationships to unfold naturally instead of forcing a direction, end goal, or time period
- Judicious and patient but not guarded
- Pursues and freely expresses intimacy, empathy, and unconditional love
- Has strong values while avoiding judging human limitations in themselves and others
- Comfortable in their own skin and being with others
- Reads situations and others and adjusts to their requirements and personalities
- Acknowledges others in ways meaningful to them
- Takes full responsibility for how they make other people feel while being true to themselves

- Known for a joyful, warm heart despite being a strong, confident, directed personality
- Sees and promotes potential in themselves and others while respecting each person's sovereignty
- Offers unconditional love to an uncommon degree
- Influential with the people they know and meet

EFFECTIVELY FRAMES AND MAKES DECISIONS

- Sees challenges, adversity, and complex decisions not as burdens, but as opportunities to apply wisdom at consequential moments
- Decisive, eager to act and be held accountable
- Frames decisions neutrally in the context of the law of polarity, finding the merit in both opposites
- Fuses intelligence, knowledge, perspective, timeless wisdom, and eternal ethics with temporal demands into effective, pragmatic solutions honoring both polarities
- Balances idealism and pragmatism
- Progressively makes more complex and difficult decisions
- Patient with a keen sense for decision ripeness and timing
- Seeks what is right, not who is right
- Searches for the best way, not one way
- Renders stunningly wise decisions most of the time, leaving many observers awestruck
- Understands decision-making limitations: doing their best with innate talents, skills, knowledge, available information, and capacity for wisdom at the decisive moment
- Accepts human fallibility and mistakes, and learns from them to avoid destructive patterns

EXPANDING, GROWING, ENERGETIC

- Possesses a growth-oriented personality seemingly without limits
- Aspires to and assimilates the most positive human characteristics
- Is accomplished, possessing the dignity of self-achievement and enlightenment from self-discovery
- Continually strives, grows, expands, experiments, explores, tests limits, and stretches their comfort zone without an identifiable stimulus
- Rose from often humble or unfavorable beginnings with a rapid, often unexplainable transformation to who they became
- Experiences luck, coincidence, timing, grace, or good fortune, as if life continually intervenes on their behalf with optimal timing
- Understands and demonstrates the art of living and advanced wisdom skills (refer to Chapters 5 through 8)
- Always open, receptive, and willing to listen
- Calm and deliberate but never inactive
- Experiences gratitude and appreciation effortlessly, even for most minor things
- Appreciates wisdom as a capacity to develop, learn, and apply while accepting that it is imperfect so that a wise person can be unwise
- Shows up all in, every day, for themselves and others
- Disciplined in the required practices to accelerate potential and wisdom

DISPLAYS CALM STRENGTH AND COURAGE

- Magnetic to others because of their thoughtfulness with humor, calm with action, and strength with intimacy
- Often revered by others who value every opportunity to be with them and learn

- Makes everyone feel seen and heard by remaining present in the moment
- Asks penetrating, insight-provoking questions
- Is a keen observer
- Angered rarely and only by the most egregious behavior
- Finds humor and applies it to enhance joyful moments or relieve stress
- Trusts the INME's surging strength to overcome the collective Ego/Shadow
- Exhibits calm confidence consistently, and is not overwhelmed by the issues at hand or passing emotions
- Feels at peace, fearless, both warrior and peacemaker
- Holds themselves and those in their orbit accountable
- Knows the best thing to do when others are lost or confused, is never morally confused or conflicted
- Reflects the truism that wisdom is easier to experience than define

FREE AND INDEPENDENT

- No tyrant owns them, including the internal tyrants of Ego, Shadow, Fear, destructive emotions, habits, judgments, or expectations of others or society
- Demonstrates life's creative potential through their uniqueness and force of personality

A NOTE ON WHAT WISDOM ISN'T

Wisdom is intention put into practice. Intention alone will not make you wise. Age doesn't make you wise, but wisdom develops in most people to some degree with age. Experience complements the capacity for wisdom in those who learn from it and intentionally change their behavior. Intellect enhances the acquisition of knowledge and its application to decision-making. However, intellect does not guarantee that someone will become wise. Intellect can hinder wisdom if the Ego becomes elitist or assumes privilege. "Street smarts" and being "life taught" are valuable alternatives or supplements to intellect. Knowledge, like intellect, helps develop the capacity for wisdom, but knowledge alone is not wisdom. It is how knowledge is applied that matters. Contemporary success measures of wealth, fame, power, appearance, and social standing do not make someone wise. They can be a barrier to becoming wise if the individual identifies with the contemporary success measure instead of the INME.

AM I UP TO WHAT IS REQUIRED?

Given your busy life, you may hesitate to take on anything new. You may feel that you need to grasp more of this book, consider alternatives, consult professionals for advice, or doubt you can do what is required. The Ego and Shadow are whispering in your brain all the reasons you can't or shouldn't pursue your potential, even as this book advises listening to the Soul function in your gut and the INME in your heart.

Allow me to reassure you. I believe in you. You have almost completed this challenging book that likely stretched you beyond your comfort zone. You resisted Ego-generated thoughts that pursuing potential and wisdom is not for you. You have been exposed to many new, learnable, and achievable concepts and practices. If you have absorbed one-third of the content and continue to use this book as a reference, you already know enough to reach your potential and become wise. What you don't know, you will discover and learn. Mastering the materials or practices won't hold you back. Your obstacle is the Ego's domination of your consciousness. Supplanting the Ego with your INME is straightforward but requires desire, commitment, discipline, and patience.

DESIRE

Displacement of the Ego's dominance by your INME is the unavoidable but achievable task of reaching your potential and becoming wise. You must passionately desire freedom from the Ego's domination that blocks your intimacy with the Soul and INME development. Many people lack the burning desire, enabling the Ego to convince them that an unconscious life is normal and all they can achieve. The seduced person may say it is too difficult to quiet their mind (though they demonstrate the ability to concentrate in other activities), or they don't have the time (though they find time for Netflix, sports, appearance, etc.). The true obstacle is domination of consciousness by their Ego and Shadow and a lack of desire to become free to be their best, become wise, and live well.

PRIORITY COMMITMENT

Achieving your potential must be your highest personal priority. Contrary to what the Ego/Shadow (and collective Ego/Shadow) may suggest, this desire is neither selfish nor all-consuming. By becoming your best self INME, you show up as your best for your health, family, career, and relationships. Making a time and effort commitment to your potential serves the highest good for everyone and everything, not just for you as an individual. Right now, allocate an hour daily to this commitment, dividing the time into five-to-fifteen-minute segments, which is about the same amount of time most people devote to grooming and much less than they devote to daily entertainment. **All-time problems are priority problems in disguise.** If you cannot commit an hour a day to making the pursuit of potential and wisdom your priority, you are unlikely to overcome the Ego's domination of your consciousness and life.

Reinforcing commitment has been effective for me and others as noted before in this book. It takes a few seconds every morning or several times a day. Look yourself in the eyes in a mirror and speak silently or ideally aloud:

"I will do my best to be my best, and leave the world and the people I meet a little better than when I found them."

These twenty-five words, said from the heart, represent nature's highest calling to all living things: to reach their potential and contribute to their species and the whole. This spoken commitment activates your inner growth mechanism, connects you to the INME and the Soul, and helps break through the Ego's distortions and illusions. All it takes is this commitment to do your best to be your best.

DISCIPLINE (TO SUPPLANT THE EGO WITH THE INME)

Learn to recognize and differentiate between the Ego/Shadow and the real you/INME. Use the Chapter 4 descriptions of the Ego, Shadow, Soul, Conscience, and INME to learn the distinctive voices, attitudes, and thought patterns of each psychological function. Recognize them in yourself and others. The Ego represents attitudes you don't like about "yourself" and usually find troublesome in others. The Shadow is the knee-jerk defensiveness that embarrasses you and the hypercritical voice that shows up when we feel vulnerable. With consistent practice, you will recognize each psychological function easily and quickly.

Learn to quiet the Ego

The real you/INME desires growth, and the Ego blocks it. Most people fall short of their potential because they surrender to Ego domination, lacking the desire, priority commitment, discipline, or patience to quiet their Ego.

Supplant the Ego with the INME

Use active imagination to visualize how you work with the psychological functions. Visualize the INME as the maestro of consciousness, with the Ego and other functions as the musical instruments. Imagine the INME as a computer operating system, swapping the other psychological functions as applications in and out of foreground and background.

Keep the INME in consciousness under stress

Decisions requiring wisdom are always challenging and involve high stakes. Tense, uncertain times will test your trust in the INME. Anticipate and be prepared to overcome Ego attempts to make you fearful of an incorrect decision, encourage you to distrust others, or become manipulative.

Patience (to connect with the Soul function)

After learning to quiet your Ego, connect with the Soul function to hear its guidance in whatever form it appears—as inspirations, ideas, aha! moments, meditation, or other. Your intention for connecting is critical. It should be the simple joy of intimacy with your Soul. You will need to be open and receptive and hope the Soul responds. Be patient. It may take weeks or months the first time. Practice feeling open and receptive without expectation. Listen and look around you for a Soul response. If you have difficulty, try the suggested practices, and repeat until there is a connection. The tension between the Soul and the Ego provides the creative energy to grow, shape the INME, and develop wisdom. Most people are unaccustomed to accessing the Soul, but once it clicks, you will know exactly what to do.

WISDOM "ASSIST" AUTOMATION

Unprecedented and previously unimaginable automation will soon be available to support you in the pursuit of your potential and wisdom. It can only assist the process, not replace the desire, commitment, discipline, and patience explained above.

You will be able to connect to vast works and practices of wisdom, from legendary philosophers to modern-day leaders and teachers, engaging in Socratic dialogue with AI avatars created from their works, and video and audio recordings of contemporary people. Applications already exist to capture and provide AI avatar playback of people's life histories for descendants, much as tribal people used oral histories for millennia to pass wisdom to younger generations. The stories and decisions will become wisdom-assist resources.

Passing life experiences and wisdom to later generations will become a societal norm as passing wisdom's value is appreciated, and the process becomes simple, cheap, and easy to use. Everyone can contribute to democratizing wisdom by passing on their life experience as examples of wise and unwise actions and decisions.

REACH YOUR POTENTIAL AND BECOME WISE

Nothing is stopping you. If you passionately desire to live your best self and best life and commit to its practices with patience, you will free yourself from the Ego's domination, operate at your potential, and wisdom will be yours. You will become a vanguard of this era's step change to democratize potential and wisdom with all its benefits to you, your loved ones, and future generations.

MAKING YOUR DECISION

Framing the decision to pursue potential and wisdom is critical. Let's examine a few approaches.

WHEN SHOULD I DECIDE?

Delaying is often a disguise for *"no."* You can decide to pursue your potential and wisdom at any point in your life, but since wisdom is the art of living well, why would you choose later instead of now? Only the Ego and Shadow would attempt to convince you to delay living well.

AVOID THE INERTIA BIAS

An inertia bias exists for several reasons, most of which stem from cognitive distortions, illusions, and fictions of the Ego: fear of making a mistake (decisions are different paths, neither wrong/right nor good/bad), fear of changing the status quo (change is constant and unavoidable), and fear of charting an unknown course with no guarantees (uncertainty is ever-present; certainty is fiction). The inaction bias increases when a moral question is involved.[2]

A practice that can serve you well is reversing the thought process when making consequential decisions. Instead of asking yourself why you should consider doing something, reverse it and ask questions like: *Are there compelling reasons not to act now? What pertinent information am I lacking? Am I letting things naturally unfold or am I procrastinating or really saying no?*

Instead of building a case to act, which triggers the inertia/inaction bias, take the opposite course to justify why you should not act. Reversing the inertia bias could apply to your decision to pursue your potential and wisdom by posing these questions to yourself:

Given the challenges of the New Reality, why would I reject adapting by achieving my potential and becoming wise since it is private, requires little time, and virtually no expense? Do I have a better adaptive approach?

A MORAL OBLIGATION

A wise friend once described "shining your light" as the most virtuous activity available to us as humans. The light is the creative spark that illuminates the INME and fuels every person. She loved flowers, vegetable gardens, and trees, believing that we share their innate drive to "fully flower" in our time on earth. She would say, "What would become of us if flowers, vegetables, and trees chose not to share their bounty because they were afraid of failing or too busy?"

We short-circuit creation's infrastructure if we fail to bloom or shine our light by doing our best, being our best, and showing the way for others. Ask yourself: *Do I have a moral responsibility to fully utilize my potential and become my best?*

THE LAW OF POLARITY AND THE TENSION OF THE OPPOSITES

Reducing all complex decisions requiring wisdom to neutral, equally valuable arguments or polarities has been emphasized. How would you frame the decision to pursue potential and wisdom as two neutral, valuable polarities? Below is one possibility. Remember that each polarity must have equal merit to maximize the tension.

We have a moral duty to maximize our potential.	<--------------->	The law of diversity creates a broad spectrum of discipline and commitment in people.

How could this polarity's tension generate a resolution for your decision? How about this: "I have a moral obligation to do my best and be my best to the limits of my ability. It is all that life can ask of me, and all I can ask of myself."

THE CIRCUMAMBULATED
JOURNEY TO THE BEST YOU

More than forty years ago, I made the decision you are now considering. I'll close this book by summarizing my experiences and those of others who shared theirs with me to help you visualize what your experience and rewards might be from achieving your potential, becoming wise, and living well.

Circumambulation is a nonlinear, organic transition from the prior Ego-dominated self to the real you/INME. As a student, you signal readiness to learn by being curious, open, receptive, and committed, listening, and performing the practices. Then, you patiently wait for the Teacher Soul to appear. Each time it does, you are rewarded with truth, perspective, or insight that strengthens your INME.

This is a wholly practical rather than theoretical process. You think more clearly and act more decisively and wisely. The development of potential and wisdom becomes cumulative, accelerating you forward in an ecstatic experience of perpetual growth. Inspirations and insights come easier and faster. Separations between previously siloed parts of your life disappear, and relationships between threads of your life become apparent. Your path to your potential is not a painful journey, as some might think, but a thrill-filled action vacation where each new insight and perspective is like discovering a new mountain or pristine alpine lake.

Every day becomes extraordinary when you look in the mirror as described above and resolve in your heart: "I will do my best, to be my best, and leave the world and the people I meet a little better than I found them." You know precisely what you must do that day. Every night is rewarding when you confirm: "I did my best today to be my best, and I left the world and the people I touched a little better." This simple psychological body wellness check certifies success, and you recommit to living another day in this way.

After many such days, the new, better real you/INME emerges, sometimes in small, progressive, unnoticeable steps and sometimes in unforgettable leaps of consciousness and insight. We realize we have become a different person while inhabiting the same body and circumstances. We would not change places with anyone in the world. We become confident, free, and fearless because we know that we can deal with whatever life serves up.

Possession loses its allure as we understand that nothing in life is owned and all is on loan. Nothing was ever ours to begin with, even our lives and loved ones. **As we release the Ego's desire for illusory possession, our relationships become deeper and richer.** Joy fills our hearts more often simply by another's presence. Our lives become far more focused and rewarding as we stop doing what our Ego or others expect and do only what inspires us, teaches us, or enhances intimacy.

The past no longer troubles us. We lose any desire to rewrite our lives or redo mistakes or decisions. Everything we did in the past was based on our knowledge and capacity for wisdom at the time and our bondage to the Ego. We could have done no more. Changing the past, even if it were possible, would bring us to a different, unknown place, and we love where we are.

We sense being part of life's rhythm rather than being baffled by it or opposing it. The previous divisions between our competing inner functions dissipate. Ego, Shadow, Soul, and Conscience work in harmony with the INME. We love life's reality as it is. Its laws, rhythms, adversities, opposites, and infinite unfolding and unknowns fuel rather than frighten. The macrocosm-microcosm moves from theory to reality as all of it becomes part of us, and we become a part of it all. Life is rich and magnificent and worthy of our deepest gratitude, perfect in its imperfections as we are, as I am.

A new, unique pain comes from this new place we live. It hurts far more today to witness the loss of human potential, whether in a young person's death or potential discarded by those blind to the light they carry inside. We accept it as part of life's rhythm, but in the hope that someday it will no longer be. The pain of lost potential and wisdom—the tension between what is and what could be—inspired me to write this book. It is also why I hope with all my heart that you choose to pursue the potential and wisdom that is within your grasp.

By doing and being our best to leave the world and those we touch a little better, we alter infinity in ways we cannot imagine. We are certain that our life matters. Our work is not done. As for me, I have not and will not quit until I can do no more. Until my last breath, there will be more to learn and do, but if this is my last, I am complete.

As I approach the end of life, I do not fear death, but I know the dying process may test my courage. My sincerest wish is to die with grace and dignity as my last gift to my loved ones. In this life, I have been everything and nothing: a unique drop in an endless sea, an imperfectly perfect grain of sand on an infinite beach, and a snowflake unique among billions on a perpetual range of snowcapped peaks designed by the greatest of all creative artists. On the wings of this truth, I will leave this existence to begin another unknown.

For those I love and those who love me, or if you have connected with me intimately through this book, I will always be with you. In our intimate connection, we were one, so now I live inside you forever. You know what I would think and do without effort, so I am always your trusted counsel. You know with certainty how deeply I love you, so you are forever loved. I will be there when you call me, so you are never alone.

This is the testament of those, me especially, who made and lived the decision you must now make. Whatever you choose, I wish you the best.

LIFE'S UNAVOIDABLE QUESTIONS

A DEEPER DIVE FOR INDIVIDUALS AND DISCUSSION GROUPS

TO BE WISE, WE MUST SEEK TRUTH AND REALITY in the outer world and our inner world. Answering foundational questions prepares you for decisions where moral clarity is essential. This book examines some of life's most important unavoidable questions and illustrates the depth of inquiry and analysis every question deserves. Use those examinations as models to identify and answer the unavoidable questions listed below or ponder some of your own.

- Does my life and how I live it matter?
- Are humans evolving to extinction or potential?
- Are people composed of mostly inferior qualities, such as evil and laziness, or admirable qualities, such as goodness and industriousness?
- Are people getting worse or better?
- What is the nature of forgiveness, and how does one know when and how to forgive?
- How does one identify and follow a meaningful passion?
- Does the technology shrinking the world advance greater commonality, diversity, or both?
- Why is life so hard and sometimes even seemingly cruel?

- Why do some people suffer more than others?
- How can I be sure that my intentions are genuine and that my actions are consistent with my intentions?
- Why do we struggle to do the right thing for ourselves, such as devoting proper attention to our health, wealth, wisdom, success, intimacy, and relationships?
- Will I adapt to the most rapid and extensive changes in history by becoming my best and wisest, or will I do nothing and endure the suffering that results?
- What does it mean to be human?
- Will working and living with humanlike automation emphasize or blur my human distinctiveness?
- How do I integrate with automation like AI and robotics without losing my humanity?
- What security, backup, and controls should I insist on as my life becomes increasingly dependent upon ever more intelligent and competent automation?
- How do I prepare for career opportunities as automation assumes the grunt work and frees me for higher-value work that is creative, problem-solving, and involves complex relationships with humans and humanlike automation?
- Will I navigate rapid change to seize opportunities or be frozen by fear and blinded by pessimism?
- If institutions continue faltering, how, when, and on what matters will I fill the void they leave, protect my interests, and participate in reshaping them?
- Will I lead wisely to heal polarizing issues in my orbit, or stay silent and hope someone else leads?
- Will I develop and execute an adaptive strategy for a longer life in an aging society, or trust that the government or someone else will take care of me and my loved ones?
- How will my loved ones and I assess the choices and consequences to make the wisest possible decisions in a personalized world of

expanding choices, individual initiative and productivity, and self-discovery?

- Given the challenges of the New Reality, why would I reject adapting by achieving my potential and becoming wise since it is private, requires little time, and virtually no expense?
- Do I have a better adaptive approach than my potential and wisdom for the New Reality that is emerging?
- Do I have a moral responsibility to fully utilize my potential and my best?
- How did humans get to where we are today?
- What can we each do to advance our civilization more rapidly and safely?

ENDNOTES

CHAPTER 1

1   Jung, Carl Gustav. *Memories, Dreams, Reflections*. Translated by Richard Winston and Clara Winston. New York: Pantheon Books, 1963.

CHAPTER 2

1   A. C. Wilson, "This Is Why You Matter: The Theory of Macrocosm & Microcosm," *Medium*, September 12, 2019, https://medium.com/@iamalexcwilson/this-is-why-you-matter-the-theory-of-macrocosm-microcosm-6f55c5e12d49.

2   NASA, "SnowEx / Snow," https://snow.nasa.gov/campaigns/snowex#:~:text=Seasonal%20snow%20cover%20is%20the,forcing%20of%20the%20Earth%27s%20climate.

3   CDC, "Suicide Data and Statistics," Centers for Disease Control and Prevention, published August 10, 2023, accessed February 13, 2024, https://www.cdc.gov/suicide/suicide-data-statistics.html.

4   J. Howard, "Plague Was One of History's Deadliest Diseases—Then We Found a Cure," *National Geographic*, published July 6, 2020, accessed February 23, 2024, https://www.nationalgeographic.com/science/article/the-plague#.

5   Peter H. Diamandis, "Why I'm Optimistic About the Future," Diamandis.com, published November 9, 2023, accessed February 13, 2024, https://www.diamandis.com/blog/scaling-abundance-series-19.

6   National Safety Council, "Car Crash Deaths and Rates," NSC, accessed February 13, 2024, https://injuryfacts.nsc.org/motor-vehicle/historical-fatality-trends/deaths-and-rates/.

7   CDC, "Leading Causes of Death," Centers for Disease Control and Prevention, published January 17, 2024, accessed February 13, 2024, https://www.cdc.gov/nchs/fastats/leading-causes-of-death.htm.

8   Marine Corps University, "What Is Locus of Control?," USMCU, accessed March 18, 2024, https://www.usmcu.edu/Portals/218/What%20is%20Locus%20of%20Control%20by%20James%20Neill.pdf.

CHAPTER 3

1   "natural law, n." OED Online, Oxford University Press, March 2023, www.oed.com/view/Entry/124924. Accessed 15 April 2023.

2   https://bmcbiol.biomedcentral.com/articles/10.1186/s12915-021-00990-w.

3   https://www.jstor.org/stable/24710645.

4   Frankl, Viktor E. *Man's Search for Meaning*. Translated by Ilse Lasch. Boston: Beacon Press, 2006.

5   "Sengcan Quotes (Author of True Heart)." Goodreads, accessed February 14, 2024, https://www.goodreads.com/author/quotes/3143635.Sengcan.

6   "Scientists, Philosophers Identify Nature's Missing Evolutionary Law," *Science Daily*, published October 16, 2023, accessed February 14, 2024, https://www.sciencedaily.com/releases/2023/10/231016163100.htm#:~:text=Summary%3A,patterning%2C%20diversity%2C%20and%20complexity.

7   A. Bloomenthal, "Bell Curve Definition: Normal Distribution Meaning Example in Finance," Investopedia, published September 7, 2023, accessed February 23, 2024, https://www.investopedia.com/terms/b/bell-curve.asp.

8   K. E. Himma, "Natural Law," Internet Encyclopedia of Philosophy, accessed February 14, 2024, https://iep.utm.edu/natlaw/.

9   https://www.worldhistory.org/article/664/aesops-fables/.

10 Andrew Rosen, "How Your Retirement Savings Compare to the National Average," *Forbes*, March 2, 2023, https://www.forbes.com/sites/andrewrosen/2023/03/02/how-your-retirement-savings-compare-to-the-national-average/?sh=3836a6236b2a.

CHAPTER 4

1   R. J. Schlegel et al., "Thine Own Self: True Self-Concept Accessibility and Meaning in Life," *Journal of Personal and Social Psychology*. 96, no. 2 (February 2009): 473–90, doi: 10.1037/a0014060. PMID: 19159144; PMCID: PMC4714566.

2   *Oxford English Dictionary*, "Conscience (n.), sense I.1.a," December 2023, https://doi.org/10.1093/OED/6524671354.

3   Encyclopædia Britannica, "Ego," January 18, 2023, accessed February 23, 2023, https://www.britannica.com/topic/ego-philosophy-and-psychology.

4   "gut feel, n." *Oxford English Dictionary* Online, Oxford University Press, March 2023, www.oed.com/view/Entry/82689.Accessed 15 April 2023.

CHAPTER 5

1   "transcendence, n." *Oxford English Dictionary* Online, Oxford University Press, March 2023, www.oed.com/view/Entry/203819. Accessed 15 April 2023.

2   N. Thompson, "Ray Kurzweil on Turing Tests, Brain Extenders, and AI Ethics," *WIRED*, published November 13, 2017, accessed February 23, 2024, https://www.wired.com/story/ray-kurzweil-on-turing-tests-brain-extenders-and-ai-ethics/.

3   "Musk's Neuralink to Start Human Trial of Brain Implant for Paralysis Patients," *Reuters*, published September 20, 2023, accessed February 23, 2024, https://www.reuters.com/technology/musks-neuralink-start-human-trials-brain-implant-2023-09-19/.

4   Roger Penrose, *The Emperor's New Mind: Concerning Computers, Minds, and the Laws of Physics* (Oxford: Oxford University Press, 1989), 580.

CHAPTER 6

1   Malcolm Gladwell, *Talking to Strangers* (New York, NY: Little, Brown and Company, 2019), 73.

2   Twist, Lynne. *The Soul of Money: Transforming Your Relationship with Money and Life*. New York: W.W. Norton & Company.

3   Sources: Based on Hayim H. Donin, *To Be a Jew: A Guide to Jewish Observance in Contemporary Life* (New York: Basic Books, 1991).

CHAPTER 8

1   "intimacy, n." *Cambridge English Dictionary*, Cambridge University Press, 2023, https://dictionary.cambridge.org/dictionary/english/intimacy. Accessed 15 April 2023.

2   "Intimacy." *Merriam-Webster Dictionary*, Merriam-Webster, 2023, https://www.merriam-webster.com/dictionary/intimacy. Accessed 15 April 2023.

3   Austen, Jane. *Sense and Sensibility*. Edited by Edward Copeland. Cambridge: Cambridge University Press, 2006.

4   "About the Correspondence between John and Abigail Adams," Massachusetts Historical Society, accessed February 23, 2024, https://www.masshist.org/digitaladams/archive/letter/.

5   *Steel Magnolias.* Directed by Herbert Ross. Los Angeles, CA: TriStar Pictures, 1989. Film.

6   Banks, Melissa. *The Girl's Guide to Hunting and Fishing.* New York: Penguin Books, 1999.

7   James Flynn, "Why Our IQ Levels Are Higher Than Our Grandparents," TED, video, https://www.ted.com/talks/james_flynn_why_our_iq_levels_are_higher_than_our_grandparents?language=en.

CHAPTER 9

1   F. D. Roosevelt, *The Public Papers and Addresses of Franklin D. Roosevelt with a Special Introduction and Explanatory Notes by President Roosevelt: Volume 2, The Year of Crisis*, 1933, S. I. Rosenman, ed. (Random House, 1938), 11–16.

2   "fear, n." *Oxford English Dictionary*, Oxford University Press, 2023, https://www.lexico.com/definition/fear. Accessed 15 April 2023.

3   Ekman, Paul. "What is Fear?" Paul Ekman Group, October 5, 2014. https://www.paulekman.com/what-is-fear/.

4   "anxiety, n." *Cambridge Dictionary*, Cambridge University Press, 2023, https://www.lexico.com/definition/anxiety. Accessed 15 April 2023.

5   "insecurity, n." *Oxford English Dictionary*, Oxford University Press, 2023, https://www.lexico.com/definition/insecurity. Accessed 15 April 2023.

6   "Insecurity." WebMD, 2023. https://www.webmd.com/mental-health/what-is-insecurity. Accessed 15 April 2023.

7   K. Albrecht, "The (Only) 5 Fears We All Share," *Psychology Today*, published February 23, 2024, https://www.psychologytoday.com/blog/brainsnacks/201203/the-only-5-fears-we-all-share.

8   University of Calgary, "Law of Conservation of Energy," Energy Education, accessed February 23, 2024, https://energyeducation.ca/encyclopedia/Law_of_conservation_of_energy.

9   K. Lotzof, "Are We Made of Stardust?," Natural History Museum, accessed February 23, 2024, https://www.nhm.ac.uk/discover/are-we-really-made-of-stardust.html#:~:text=Planetary%20scientist%20and%20stardust%20expert,%27.

CHAPTER 10

1   Brach, Tara. *Radical Acceptance: Embracing Your Life with the Heart of a Buddha.* New York: Bantam, 2004.

2   Frankl, Viktor E. *Man's Search for Meaning.* Translated by Ilse Lasch. Boston: Beacon Press, 2006.

3   Hume, David. "Of the Origin of Justice and Property." In A Treatise of Human Nature. Edited by L. A. Selby-Bigge. Oxford: Clarendon Press, 1896.

4   N. Burton, "The Psychology of Gratitude," *Psychology Today*, September 23, 2014, https://www.psychologytoday.com/us/blog/hide-and-seek/201409/the-psychology-gratitude.

5   F. Franco, "The Chronically Dissatisfied: Making the Connection Between Gratitude and Well-Being," PsychCentral, November 10, 2017, https://psychcentral.com/lib/the-chronically-dissatisfied-making-the-connection-between-gratitude-and-well-being#1.

6   N. Burton, "The Psychology of Gratitude."

7   R. Yehuda et al., "The Memory Paradox," *Nature Reviews Neuroscience* 11, no. 12 (December 2010): 837–839. doi: 10.1038/nrn2957. PMID: 21088685.

8   Knight, Phil. *Shoe Dog: A Memoir by the Creator of Nike.* New York: Simon & Schuster, 2016.

9   "Crucible." *Merriam-Webster Dictionary,* Merriam-Webster, 2023, https://www.merriam-webster.com/dictionary/crucible. Accessed 15 April 2023.

10 Encyclopædia Britannica, "Persona," April 4, 2008, accessed March 4, 2024, https://www.britannica.com/science/persona-psychology.

CHAPTER 11

1   D. A. Fischer, "Wisdom—The Answer to all the Questions Really Worth Asking," *International Journal of Humanities and Social Science* 5, no. 9 (2015), https://www.psychologie.uni-heidelberg.de/ae/allg/mitarb/af/7b.pdf.

2   N. Barak-Corren and M. H. Bazerman, "Inaction and Decision Making in Moral Conflicts," *Organizational Dynamics* 49, no. 1 (2020), https://doi.org/10.1016/j.orgdyn.2019.02.005.

BIBLIOGRAPHY

Alexander, Eben. *Proof of Heaven: A Neurosurgeon's Journey into the Afterlife.* Kindle ed., Simon & Schuster, 2012.

Anderson, Chris. *The Long Tail: Why the Future of Business Is Selling Less of More.* Hyperion, 2006.

Andrews, Andy. *The Traveler's Gift: Seven Decisions That Determine Personal Success.* Kindle ed., Thomas Nelson, 2002.

Baker, Dan, Cathy Greenberg, and Collins Hemingway. *What Happy Companies Know: How the New Science of Happiness Can Change Your Company for the Better.* Pearson Prentice Hall, 2006.

Ballard, Zari. *When Love Is a Lie: Narcissistic Partners & the Pathological Relationship.* Kindle ed., Zari Ballard, 2016.

Berger, Alex. *Practical Curiosity: The Guide to Life, Love & Travel.* Kindle ed., 2017.

Blackmore, Susan J. *Consciousness: A Very Short Introduction.* Kindle ed., OUP Oxford, 2017.

Brooks, Arthur C. *Gross National Happiness Why Happiness Matters for America--and How We Can Get More of It.* Basic Books, 2008.

Brown, Brené. *Daring Greatly: How the Courage to Be Vulnerable Transforms the Way We Live, Love, Parent, and Lead.* Kindle ed., Avery, 2012.

Brown, Brené. *The Gifts of Imperfection: Let Go of Who You Think You're Supposed to Be and Embrace Who You Are.* Kindle ed., Hazelden, 2010.

Browne, John. *Make, Think, Imagine: Engineering the Future of Civilization.* New York, NY: Pegasus Books Ltd, 2019.

Browne, John. *Make, Think, Imagine.* Pegasus Books, 2019.

Cappannelli, George C., and Sedena C. Cappannelli. *Do Not Go Quietly: A Guide to Living Consciously and Aging Wisely for People Who Weren't Born Yesterday.* Agape Media International, LLC, 2014.

Carey, Benedict. *How We Learn: The Surprising Truth About When, Where, and Why It Happens*. Kindle ed., Random House, 2014.

Carnegie, Dale. *How to Win Friends and Influence People*. Kindle ed., Simon & Schuster, 2022.

Chapman, Gary D. *The 5 Love Languages: The Secret to Love That Lasts*. Kindle ed., Northfield Publishing, 2014.

Charvet, Shelle Rose. *Words That Change Minds: The 14 Patterns for Mastering the Language of Influence*. Kindle ed., Institute for Influence, 2019.

Christensen, Clayton M., and Henry J. Eyring. *The Innovative University: Changing the DNA of Higher Education from the Inside Out*. Kindle ed., Jossey-Bass, 2011.

Christensen, Henriette Eiby. *110 Ways to Spot a Toxic Person: I Love No One*. Kindle ed., CreateSpace Independent Publishing Platform, 2013.

Cline, Eric H. *1177 B.C.: The Year Civilization Collapsed*. Kindle ed., Princeton University Press, 2021.

Coelho, Paulo. *The Alchemist*. 25th Anniversary ed., HarperOne, 2015. Kindle ed.

Conley, Chip. *Wisdom at Work: The Making of a Modern Elder*. Kindle ed., Crown Currency, 2018.

Coughlin, Joseph F. *The Longevity Economy: Unlocking the World's Fastest-Growing, Most Misunderstood Market*. New York, NY: PublicAffairs, 2017.

Diamond, Jared. *Guns, Germs, and Steel: The Fates of Human Societies*. New York, NY: W. W. Norton & Company, 1999.

Dispenza, Joe. *Breaking the Habit of Being Yourself*. Kindle ed., Hay House LLC, 2012.

Dispenza, Joe. *You Are the Placebo: Making Your Mind Matter*. Kindle ed., Hay House LLC, 2014.

Easterbrook, Gregg. *It's Better Than It Looks: Reasons for Optimism in an Age of Fear*. Kindle ed., PublicAffairs, 2018.

Finlayson, Clive. *The Humans Who Went Extinct: Why Neanderthals Died out and We Survived*. Kindle ed., OUP Oxford, 2009.

Frankl, Viktor E. *Man's Search for Meaning*. Kindle ed., Beacon Press, 2006.

Garcia, Manny. *A Glossary of Life: Deeper Meaning behind Our Common Words*. Manny Garcia, 2018.

Gawande, Atul. *Being Mortal: Medicine and What Matters in the End*. Kindle ed., Metropolitan Books, 2014.

Gilbert, Elizabeth. *Big Magic: Creative Living Beyond Fear*. Kindle ed., Riverhead Books, 2015.

Gingrich, Newt, Vince Haley, and Rick Tyler. *Real Change: From the World That Fails to the World That Works*. Regnery Publishing,Inc., 2007.

Diamond, Jared. *Guns, Germs, and Steel: The Fates of Human Societies*. New York, NY: W. W. Norton & Company, 1999.

Gingrich, Newt. *Winning the Future: A 21st Century Contract with America*. Regnery Publishing Inc., 2005.

Gladwell, Malcolm. *Talking to Strangers: What We Should Know about the People We Don't Know*. Kindle ed., Little, Brown and Company, 2019.

Gladwell, Malcolm. *The Tipping Point: How Little Things Can Make a Big Difference*. Little, Brown and Company, 2000.

Goleman, Daniel. *A Force for Good: The Dalai Lama's Vision for Our World*. Kindle ed., Bantam, 2015.

Goleman, Daniel. *Emotional Intelligence: Why It Can Matter More Than IQ*. Kindle ed., Bantam, 2012.

Harling, Becky. *How to Listen So People Will Talk: Build Stronger Communication and Deeper Connections*. Kindle ed., Bethany House Publishers, 2017.

Harris, Annaka. *Conscious: A Brief Guide to the Fundamental Mystery of the Mind*. Kindle ed., Harper, 2019.

Harvard Business Review Entrepreneur's Handbook: Everything You Need to Launch and Grow Your New Business. Harvard Business Review Press, 2018.

Heikura, Elisa. "Know the Difference between Shame, Guilt, Humiliation and Embarrassment." *Developerhood*, March 27, 2020. https://developerhood.com/blog/do-you-know-the-difference-between-shame-guilt-humiliation-and-embarrassment/.

Hendrix, Harville, and Helen Hunt. *Getting the Love You Want: A Guide for Couples*. Kindle ed., St. Martin's Griffin, 2019.

Holiday, Ryan. *Ego Is the Enemy*. Kindle ed., Portfolio, 2016.

Howes, Lewis. *The Mask of Masculinity: How Men Can Embrace Vulnerability, Create Strong Relationships, and Live Their Fullest Lives*. Kindle ed., 2017.

Isay, Jane. *Walking on Eggshells: Navigating the Delicate Relationship Between Adult Children and Parents*. Kindle ed., Anchor Books, 2008.

Jacoby, Tamar, ed. *This Way Up: New Thinking About Poverty and Economic Mobility*. American Enterprise Institute, 2016. PDF. https://media4.manhattan-institute.org/sites/default/files/This-Way-Up-0118.pdf.

Jamison, Kay Redfield. *An Unquiet Mind: A Memoir of Moods and Madness*. Kindle ed., Vintage, 2009.

Johnson, Larry, and Bob Phillips. *Absolute Honesty: Building a Corporate Culture That Values Straight Talk and Rewards Integrity*. American Management Association, 2003.

Johnson, Paul. *Socrates: A Man for Our Times*. Kindle ed., Penguin Books, 2011.

Jones, Judy, and William Wilson. *An Incomplete Education: 3,684 Things You Should Have Learned but Probably Didn't*. Kindle ed., Ballantine Books, 2009.

Kabat-Zinn, Jon. *Wherever You Go, There You Are: Mindfulness Meditation in Everyday Life*. Kindle ed., Hachette Go, 2023.

Kidder, Tracy. *Strength in What Remains: A Journey of Remembrance and Forgiveness.* Kindle ed., Random House, 2009.

Kotler, Steven. *Stealing Fire: How Silicon Valley, the Navy SEALS, and Maverick Scientists Are Revolutionizing the Way We Live and Work.* Audiobook ed., HarperAudio, 2017.

Kotler, Steven. *The Rise of Superman: Decoding the Science of Ultimate Human Performance.* Kindle ed., Amazon Publishing, 2014.

Kurzweil, Ray. *The Singularity Is Near: When Humans Transcend Biology.* Kindle ed., Penguin Books, 2005.

Leutzinger, Joseph, and John Harris. *Why and How People Change Health Behaviors.* Health Improvement Solutions Inc., 2005.

Levy, Ariel. *The Rules Do Not Apply: A Memoir.* Kindle ed., Random House, 2017.

Lue, Natalie. *The No Contact Rule.* Kindle ed., Naughty Girl Media, 2013.

Lyubomirsky, Sonja. *The Myths of Happiness: What Should Make You Happy but Doesn't, What Shouldn't Make You Happy but Does.* Kindle ed., Penguin, 2013.

MacCulloch, Diarmaid. *Christianity: The First Three Thousand Years.* Kindle ed., Penguin Books, 2010.

Mason, Paul. *Stop Walking on Eggshells: Taking Your Life Back When Someone You Care About Has Borderline Personality Disorder.* Kindle ed., New Harbinger Publications, 2020.

Maurer, Michael S. *10 Essential Principles of Entrepreneurship You Never Learned in School.* IBJ Book Pub LLC, 2012.

McRaven, William H. *Make Your Bed: Little Things That Can Change Your Life ... and Maybe the World.* Kindle ed., Grand Central Publishing, 2017.

Moore, Thomas. *A Religion of One's Own: A Guide to Creating a Personal Spirituality in a Secular World.* Kindle ed., Avery, 2014.

Mullainathan, Sendhil, and Eldar Shafir. *Scarcity: Why Having Too Little Means So Much.* Kindle ed., Times Books, 2013.

Mayo Clinic. *Mayo Clinic Book of Alternative Medicine.* Time Inc. Books, 2013.

Nesse, Randolph M., and George C. Williams. *Why We Get Sick: The New Science of Darwinian Medicine.* New York: Times Book, Inc., 1995..

Nichols, Wallace J. *Blue Mind: The Surprising Science That Shows How Being In, On or Under Water Can Make You Happier, Healthier, More Connected and Better at What You Do.* Kindle ed., Little, Brown and Company, 2014.

Osho. *Intimacy: Trusting Oneself and the Other.* Kindle ed., St. Martin's Griffin, 2007.

Pagán, Camille. *Life and Other Near-Death Experiences.* Kindle ed., Lake Union Publishing, 2015.

Pausch, Randy, and Jeffrey Zaslow. *The Last Lecture.* Kindle ed., Hachette Books, 2008.

Peters, Thies. *Liberation Management: Necessary Disorganization for the Nanosecond Nineties.* A.A. Knopf, 1992.

Peters, Thomas J., and Nancy J. Austin. *A Passion for Excellence: The Leadership Difference.* Profile Books Limited, 1994.

Peters, Thomas J. *The Tom Peters Seminar: Crazy Times Call for Crazy Organizations.* Knopf Doubleday Publishing Group, 2010.

Peters, Tom H., and Robert H. Waterman. *In Search of Excellence: Lessons from America's Best-Run Companies.* HarperCollins Publishers, 2006.

Pink, Daniel H. *When: The Scientific Secrets of Perfect Timing.* Kindle ed., Riverhead Books, 2018.

Pressfield, Steven. *The Artist's Journey: The Wake of the Hero's Journey and the Lifelong Pursuit of Meaning.* Kindle ed., Black Irish Entertainment LLC., 2018.

Pressfield, Steven. *The War of Art.* Kindle ed., Black Irish Entertainment LLC, 2011..

Ridley, Matt. *The Red Queen: Sex and the Evolution of Human Nature.* Kindle ed., 2nd ed., Harper Perennial, 2012.

Ridley, Matt. *The Rational Optimist: How Prosperity Evolves.* Kindle ed., HarperCollins e-books, 2010.

Robinson, Julia. *On Intimacy: A Forgotten Art.* Independently published, 2017.

Rosling, Hans, Ola Rosling, and Anna Rosling Rönnlund. *Factfulness: Ten Reasons We're Wrong About the World - and Why Things Are Better than You Think.* Flatiron Books, 2018.

Rothschild, Babette. *8 Keys to Safe Trauma Recovery: Take-Charge Strategies to Empower Your Healing.* Kindle ed., W.W. Norton & Co., 2010.

Rutherford, Adam, and Siddhartha Mukherjee. *A Brief History of Everyone Who Ever Lived: The Human Story Retold through Our Genes.* Kindle ed., The Experiment, LLC, 2018.

Sakugawa, Yumi. *I Think I Am in Friend-Love with You.* Kindle ed., Adams Media, 2013.

Schucman, Helen, and Robert Perry. *A Course in Miracles: Complete and Annotated Edition.* Kindle ed., Circle of Atonement, 2017.

Schwab, Klaus. *The Fourth Industrial Revolution.* World Economic Forum, 2016.

Sedler, Michael D. *When to Speak Up & When to Shut Up.* Kindle ed., Revell, 2006.

Soforic, John. *The Wealthy Gardener: Life Lessons on Prosperity between Father and Son.* Kindle ed., EGH Publishing, 2018.

Sternberg, Esther M. *The Balance within: The Science Connecting Health and Emotions.* Henry Holt and Company, 2001.

Strayed, Cheryl. *Brave Enough.* Kindle ed., Knopf, 2015.

Strudwick, Sarah. *Dark Souls: Healing and Recovering from Toxic Relationships.* 1st ed., S S Products, 2010.

Swithin, Tina. *Divorcing a Narcissist: Advice from the Battlefield.* Kindle ed., 2014.

Tapscott, Don, and Anthony D. Williams. *Wikinomics: How Mass Collaboration Changes Everything.* New York: Portfolio, 2006.

Thomas, M. E. *Confessions of a Sociopath: A Life Spent Hiding in Plain Sight.* Kindle ed., Reprint ed., Crown, 2013.

Toffler, Alvin. *The Third Wave.* New York: Bantam Books, 1980.

Toffler, Alvin. *Future Shock.* New York: Random House, 1970.

Toffler, Alvin. *Powershift: Knowledge, Wealth, and Violence at the Edge of the 21st Century.* Bantam Books, 1990.

Tramuto, Donato, and Chris Black. *Life's Bulldozer Moments: How Adversity Leads to Success in Life and Business.* Hamilton Books, 2016.

Tsabari, Shefali. *The Awakened Family: How to Raise Empowered, Resilient, and Conscious Children.* Kindle ed., Reprint ed., Penguin Books, 2016.

Twist, Lynne. *The Soul of Money: Transforming Your Relationship with Money and Life.* Kindle ed., W.W Norton & Company, 2010.

Vanier, Jean. *Becoming Human.* Kindle ed., Mahwah, NJ: Paulist PressTM, 2014.

Walker, Matthew P. *Why We Sleep: Unlocking the Power of Sleep and Dreams.* Kindle ed., Illustrated ed., Scribner, 2017.

Ware, Bronnie. *The Top Five Regrets of the Dying: A Life Transformed by the Dearly Departing.* Kindle ed., Hay House LLC, 2019.

Wasserman, Noam. *Life Is a Startup: What Founders Can Teach Us about Making Choices and Managing Change.* Stanford, CA: Stanford Business Books, an imprint of Stanford University Press, 2018.

Wattenberg, Ben J. *Fewer: How the New Demography of Depopulation Will Shape Our Future.* Chicago: Ivan R. Dee, 2004.

Welwood, John. *Perfect Love, Imperfect Relationships: Healing the Wound of the Heart.* Kindle ed., Boston, Mass: Trumpeter, 2011.

Williamson, Marianne. *A Return to Love: Reflections on the Principles of A Course in Miracles.* Kindle ed., Abridged ed., HarperOne, 2009.

Winfrey, Oprah. *The Path Made Clear: Discovering Your Life's Direction and Purpose.* Kindle ed., Flatiron Books, 2019.

Winfrey, Oprah. *The Wisdom of Sundays: Life-Changing Insights from Super Soul Conversations.* Kindle ed., Flatiron Books, 2017.

Wright, Robert. *The Evolution of God.* Kindle ed., 1st ed., Little, Brown and Company, 2009.

Yates, John, Matthew Immergut, and Jeremy Graves. *The Mind Illuminated: A Complete Meditation Guide Integrating Buddhist Wisdom and Brain Science for Greater Mindfulness.* Kindle ed., Atria Books, 2024.

Yogi, Mahesh. *The Science of Being and Art of Living.* Kindle ed., MUM Press, 2011.

Young, Indi. "Where Do Ideas Come From?" Medium. Inclusive
    Software, July 2, 2019. https://medium.com/inclusive-software/
    where-do-ideas-come-from-5065d685634c.

Zukav, Gary. *The Seat of the Soul: 25th Anniversary Edition*. Kindle ed., Anniversary
    ed., Simon & Schuster, 2007.

ACKNOWLEDGMENTS

WHEN AUTHORS WRITE ACKNOWLEDGMENTS for inclusion in a book, they recognize the people who contributed to its creation and usually go to lengths to point out that without those people, the book would not have happened.

The Potentialist Series and this book, *The Potentialist: The Pursuit of Wisdom*, are unusual books about subjects most people prefer to avoid—the future, whether they are living up to their innate potential, and whether they live wisely and well. To speak with authority about these subjects took a lifetime of preparation and encouragement.

The people who made these books possible co-created me. They encouraged my research and growth and expanded my capacity for wisdom throughout my long life, beginning at age twenty, when I committed to becoming my best, whatever that might be. At a minimum, their contributions have been fifty-seven years and counting. There have been thousands of such people. Some contributed longer or more intensely, but all of their influence was profound. Not one person in those thousands asked for anything in return.

It is impossible for me to write a traditional acknowledgment for this book. Instead, I thank life's creator (whatever you believe it to be) for designing and installing in each of us an immaculate system that makes possible the achievement of our innate ability. And I thank all of you selfless, wise people who co-created me by trusting in that creator's intention and handiwork. You shaped me into someone willing and able to write this book series. Through that action, you have hopefully empowered many to realize their potential and wisdom and altered their life's course and infinity.

—Ben Lytle

ABOUT THE AUTHOR

BEN LYTLE is a self-made serial entrepreneur and CEO known for being ahead of the curve. He is the author of *The Potentialist: Your Future in the New Reality of the Next Thirty Years,* a guidebook for success during the fast-changing, turbulent, and opportunity-rich times ahead.

Ben is best known as the founding CEO of Anthem, Inc. (NYSE), one of the leading US health plans with a market capitalization placing it in the top 30 of the Fortune 500; and Acordia, Inc. (NYSE), which became the world's sixth-largest insurance broker. He cofounded three companies with his entrepreneurial son, Hugh, and invests in technologies that address New Reality challenges, such as increasing productivity to offset declining populations, caring for the elderly, and preserving human legacies.

Ben has extensive public policy experience at the state and federal levels and has held board leadership roles in a wide range of industries. His contributions have been recognized by numerous awards and in books and periodicals. He has been a speaker and university guest lecturer on health, healthcare policy, entrepreneurship, and human potential throughout his career.

Beyond his career, Ben's passions include his family of three adult children and eight adult grandchildren, along with lifelong avocations for physical fitness, travel, reading, human potential, the future, and education.

INDEX

Note: Page numbers in **bold** refer to tables.

technology, 8
character, 43, 125–126, 134
    adversity and, 218
    avoidance, 133
    character-defining moral choice, 59
charitable intention, 127–128
Chesterton, G. K., 224
childhood personality traits, 92
Chopra, Deepak, 118
circumambulation, 14–15, 255–257
civilization, 28, 40, 48
"close-knit families," 168
cloud, 24
co-creation/co-creators, 13, 219–220
    destructive, 220
    family, 223
    interim co-creators, 220–221
    involuntary relationships, 221
    practices, 222
    symbiotic relationships, 220
    wise co-creators, 221–222
cognitive behavioral therapy, 19
cognitive distortions, 19, 46, 202. See also distortions
    dichotomous, 46
    as opposites of acceptance, 214–215
    of reality, 46
    self-destructive, 78
cognitive distortions of Ego, 19, 202, 214
    inertia bias, 253
    of reality, 46
    self-destructive, 78
collective Ego, 19, 20, 48, 75, 76, 81, 84, 113, 190, 214, 241, 249
collective human potential, 66
colloquialisms, 41, 42
    "go with the flow," 51
Community Antenna Television (CATV), 38
complex natural systems, 52
conditional love, 170. See also unconditional love
connectedness, 110
    reality, 24–25
Conscience, 36, 64, 68, 101, 125, 250
    definition, 73

engaging psychological functions, 208
function and experience, 74
functions, 117
INME influences, 89–90, 153, 181, 256
origin of, 69
practice identifying, 95
working with instead of against, 74
conscious intention, 117
    aligning intentions, 131
    aligning interests, 137–138
    assessment for relationships, 133
    avoiding character, brand, and relationship catastrophes, 133
    case studies, 135–136
    clarification, 132
    detection in others, 134
    making habitual, 132–133
    reinforcing growth to potential, 134
    series of choices, 130
    transparency, 132
    unappreciated impact of, 119
consciousness, 101–102, 209, 256. See also unconscious
    calm courage, 110
    and capacity for wisdom, 101
    capturing everything from soul function, 113
    as committed priority, 111
    connectedness, 110
    Conscience and, 73
    cultivating and maximizing, 102, 107–110
    differences from traditional learning, 105–106
    domination of, 65, 78, 113, 241, 248, 249
    Ego's dominancy in, 66, 78, 231, 241–242, 248–249
    embracing dynamism, 111–112
    expansion of, 109
    expecting collective ego to oppose, 113–114
    experiencing Soul's ubiquity, 109
    human's capacity for, 190
    increased intimacy with soul, 110
    INME in, 231, 250–251

Dickinson, Don, 185
dignity of self-achievement, 70, 246, 257
discipline, 250–251
dishonorable intention, 120, 134
disruption, 19, 238
distortions, 25, 202. *See also* cognitive
distortions; Ego(s)
    consequences, 80
    of perfection, 126, 150
    of reality, 19
distress, 19, 23000
diversity, 59. *See also* Law of Diversity
    as business strategy, 58
    careers and, 57
domination
    of consciousness, 65, 78, 113, 241,
    248, 249
    of environment, 72, 78
    of people, 78
Dyke, Wade, 16
dystopian thinking, 79, 151

E

educational mission, 8–9
Ego-death, 199, 201
Ego-driven thinking, 169–170
Ego(s), 19, 36, 64, 71, 101, 125, 250
    calming, 154–156, 158, 180
    cognitive distortions of, 19, 46, 78,
    202, 214, 253
    collective, 19, 20, 48, 75, 76, 81, 84,
    113, 190, 214, 241, 249
    confusion about, 75–76
    definition, 75
    dominancy in consciousness,
    65–66, 78, 108, 231, 241–242,
    248–249
    engaging psychological functions,
    208
    equating safety with success, **79**
    extreme Ego experience, 78–79
    fanaticism derives from, 48
    frames, 23
    functions, 117
    identification by consistency, 76
    INME influences, 89–90, 153, 256

    and intimacy, 172–173
    jealous, 111
    to oppose consciousness, 113–114
    practice identifying, 95
    prime mission of defense, 231–232
    principles for substituting INME
    for, 80–81
    prioritizing distortions and
    temporal success, 80
    psychological body function, 67
    quieting, 94–95, 141, 250
    self realization about, 76
    shifting consciousness to INME,
    81–82
    and solitude, 143
    Soul functions compensating
    opposition to, 85
    supplant Ego with INME, 250–251
    survival *vs.* aspirational
    characteristics, **77**
    threatened, 232
Einstein, Albert, 41, 101, 145, 227
Eisenhower, Dwight David, 122
Ekman, Paul, 192
electronic communications, 120
    cause and effect in, 121–122
emotional reasoning, 78
empathy, 57, 90, 152, 183
    and intimacy, 164, 166–168, 244
    mutual, 58
    Soul and, 85
endowments, 137
envy, 50, 80
eternal world, living in and reconciling,
67–69
experience-sharing technology, 184
extinction beliefs, 26
extinction fanaticism, 26
extinction mindset, 27
    consequences of, 26–27
    phenomenon, 26
extinction threats, 26, 30, 31
extreme Ego experience, 78–79
extreme polarization, 47
extremism, 48–49, 147

I

Ignatius, Saint, 12, 140
illogical conclusions. *See* magical conclusions
illusions
    of certainty, 203, 204
    of control, 55
    death as, 204–205
    fear based on, 202–205
    and humanity, 202
    isolation, 24–25
    as opposites of acceptance, 214–215
    of reality, 19
    reality beating, 13
impaired cognitive function, 66
impatience, 78, 211
impersonal interactions, 181–182
inconsistency, convenient, 56
individual responsibility, 47–49
Industrial Age, 3, 238
inertia bias, 253
inferior qualities, 32
ingratitude, 148, 224–225
INME ("I" in "Me"), 11, 14–15, 36, 64–65, 69–70, 206, 231
    arrival at potential, 91
    being good parent to develop, 207–208
    in consciousness, 231, 250–251
    in consciousness under stress, 251
    definition, 88
    development, 108–109
    expanding capacity for wisdom, 90
    function and experience, 89
    influences ego, Shadow, Soul, and conscience, 89–90
    intention in development, 125
    practice identifying, 95
    principles for substituting INME for ego, 80–81
    proxies, 90
    recognizing, 66, 89
    shifting consciousness from ego to, 81–82
    supplant Ego with, 250–251
    upgrading, 104–105

innate growth mechanism, 11, 12, 141
innovations, 51, 141, 147, 184, 204, 218
    AI and robotics, 8
    brain-to-computer interfaces, 8
    democratizing, 7
    quantum computing, 8
insecurity, 193, 197
instinctive Shadow, 71
"intention-accountability-best effort" learning dynamic, 126
intentions, 117–119. *See also* conscious intention
    and actions in character, brand, and relationships, 125–126
    and aligned interests, 122–125
    aligning, 131
    and capacity for wisdom, 125
    case study, 128–129
    cause and effect in electronic communications, 121–122
    challenges to social norms of presumed good intention, 120
    clarifying, 132
    detecting intention in others, 134
    dishonorable, 120
    in INME development, 125
    in leadership, 135–136
    practices, 130–133
    subtleties, 127–128
    transparency of intention, 132
    in twenty-first century, 120–122
interim co-creators, 220–221
internet cloud-based applications, 24
Intimacy, 14, 50, 87, 141, 148, 152, 163, 165
    allure, 171–172
    with animal s and inanimate objects, 172
    barriers to, 172–174
    careers and laws of, 57
    and conditional love, 170
    and consumption, 169–170
    contrasts, 166–172
    distinct and unique change, 175–176
    intimate experience, 174–179
    with oneself, 177
    opportunities, 179–180

Levine, Tim, 119
life, 19
    detecting and answering
    unavoidable questions, 37
    devaluing, 55
    and ego, 79
    expectancy, 6, 54
    foundational issues and attitudes,
    20
    gift of life and macrocosm-
    microcosm, 22–23
    humans evolving to extinction or
    potential, 25–31
    importance of, 21
    inferior qualities, 32–34
    isolation illusion and
    connectedness reality, 24–25
    life-altering change, 6
    life-changing benefits, 20
    macrocosm-microcosm in math,
    science, and technology, 23–24
    people under stress, 22
    unavoidable questions, 259–261
likelihood, 201
Lilly Endowment, 137–138
Lincoln, Abraham, 36
literacy, 6
living for oneself, 174
living for others, 174
loneliness
    devoid of, 143
    failed attempts at intimacy, 182
    threat to mental and physical
    health, 184
lost relationships, 80
love, 164
    aura of, 176
    conditional, 170
    Ego and, 169–170
    intimacy and, 50, 164
    life, 243, 256
    romantic, 171
    Soul aspiration for, 85
    spectrums of, 167–168
    unconditional, 83, 90, 168–169
low-scoring threat, 201
Lyon, Dennis, 185–186

M

macrocosm-microcosm, 25
    gift of life and, 22–23
    isolation illusion and
    connectedness reality, 24–25
    in math, science, and technology,
    23–24
    theory, 163
magical conclusions, 78
Mahatma Gandhi. *See* Gandhi,
Mohandas Karamchand
Man's Search for Meaning (Frankl), 46,
218
Maraboli, Steve, 165
mass media, 171, 190, 199, 238
mathematics, 22–23
    macrocosm-microcosm in, 23–24
meaning
    Ego's equate safety with success
    instead of, **79**
    interpreting meaning of intimate
    moments, 182–183
    life, 22, 46
    prioritizing distortions and
    temporal success over, 80
medical innovations, 8
medical science, 205
memory, 142
    Gratitude, 148
    muscle, 33, 142, 154, 215
    and past, 225–227
    state of Neutrality to, 147
    of the heart, 224
mentors, 90, 112, 126, 177, 220
mindfulness, 64
mind-reading, 78
misaligned intention, case study of,
123–125
misalignment, 122
moral obligation, 205, 254
multiple personas, 231
Musk, Elon, 105
mutilation, 198
Mutz, John, 137–138
Myers-Briggs Personality Inventory, 93

tough decisions in turbulent time, 238–240
universal ethos, 14
user's manual for real, 15
visualizing to realizing, 12
vows and pledges from heart, 12
way to reaching, 11
wind beneath wings, 13
Potentialist book series, 56, 64
Powell, Joseph, 41
power, 78
    of human spirit, 218
    of intimacy, 166
    of Souls, 178
pragmatic optimism, 30
predators, 27, 35, 206
pridefulness, 114
primal fears, 198
priority commitment, 249–250
proverbs, 42–43, 53
psychoanalytic theory, Ego in, 75
psychological body, 14, 68
    checkpoint, 69–70
    components of, 101
    conscience, 73–74
    conversations with psychological functions, 95
    discovering baseline, 92
    discovering before age six, 92
    ego, 75–82
    evolution of, 64–65
    INME, 88–91
    knowledge gap, 65–67
    living in and reconciling temporal and eternal worlds, 67–69
    optimization, 63–64
    personality predispositions and "gut feel" decision-making, 93
    physical and psychological body knowledge gap, 65–67
    practices to use, 91
    present personality, 93–94
    psychological functions overview, 70
    quieting ego, 94–95
    Shadow, 70–73
    small voice, 95–97
    Soul, 82–88

psychological functions, 70, 95, 166, 208
psychological states, 139
    computers and smartphone apps, 153
    in group, 158
    identifying, 156
    learning from unpleasant, 153–154
    selecting and swapping, 152–154
    unconscious and haphazard, 152–153
psychological theories, 14
psychology, 53
public persona, identification with, 80
puppy love, 164
Pygmalion effect, 140

Q

quantum computing, 8, 23
quantum physics, 202
quantum theory, 23

R

radical acceptance, 214
readiness, 106, 112, 255
reality, 31, 214
    accepting, 38, 214
    Ego's cognitive distortions of, 46
    of life, 90, 110
    searches for, 242–243
    seeing and, 38
reciprocity, expectation of, 174
Reflection, 141, 150–151, **157**
relationships, 87, 122, 125–126, 164
    acceptance, impact of, 213
    assessing choose for, 133–134
    assessing value of, 222
    with fear, 191–192
    fear of change impact, 203
    human relationships with automation, 9
    individual capacity for, 167
    intimacy in, 164–165, 166, 181, 244–245
    involuntary, 221

Special Theory of Relativity, 50
specific fears, 197
specificity, 201
starburst, 109
Steel Magnolias (Roberts), 173
stewardship, 203
Stone, Clement, 194
stress
    INME in consciousness under, 251
    people under, 22
*A Sunday Afternoon on the Island of La
Grande Jatte* (Seurat), 149
suspicion, 78, 119, 168
Swanson, Mary, 58
symbiotic relationships, 220

T

Talking to Strangers (Gladwell), 119
TDT. *See* truth-default theory
teachers, 220
    constructive, 72
    traditional learning depends on,
    106
technology, 8, 35
    macrocosm-microcosm in, 23–24
    technology-fueled democratization,
    7
temporal learning, 103–107
temporal world, living in and
reconciling, 67–69
tension of opposites. *See* law of polarity
Teresa, Mother, 36, 95, 96
threat assessment, 201–202
threatened ego, 232
time/timing, 50, 207
    becoming wise in foolish, 10
    and effort commitment to
    potential, 249
    humanity freeing from historic
    limitations of, 24
    between INME and Ego, 232
    physical intimacy and, 171, 178
    quantum applications and, 23
    reality beating illusions, 13
    relationships decisions and, 219
    Soul and, 84, 88, 109, 112

spending time with people, 223
    in temporal world, **68**
    tough decisions in turbulent time,
    238–240
    zones, 120
time-worn truths, 87
Toffler, Alvin, 136
traditional learning
    brevity and retention, 106
    differences from, 105
    learning pace, 105
    learning sought and learning
    delivered, 106
    pedagogy, 105
    readiness to learn, 106
trained therapists, 221
transcendence, 104
transcendent experiences, 113
transparency of intention, 132
truth-default theory (TDT), 119
Twist, Lynne, 127

U

unavoidable questions, 20
    detecting and answering, 37
    in intention, 128
    of life, 19–38, 209, 217, 259–261
    tension and, 45
    in tough decisions, 238–240
unconditional love, 83, 164, 169
    Ego and, 75, 172
    family manifestation of, 223
    INME and, 90
    intimacy and, 168–169
unconscious. *See also* consciousness
    behaviors, 71
    choices, 183
    fear, 193, 195
    intentions, 117, 118, 126, 129
    life, 249
    self-assessment, 35–36
undemocratic systems, 48
undisclosed intention, case study of,
123–125
unexpected breakthroughs, 87
unifying experiences, 87